GRIZZLY WEST

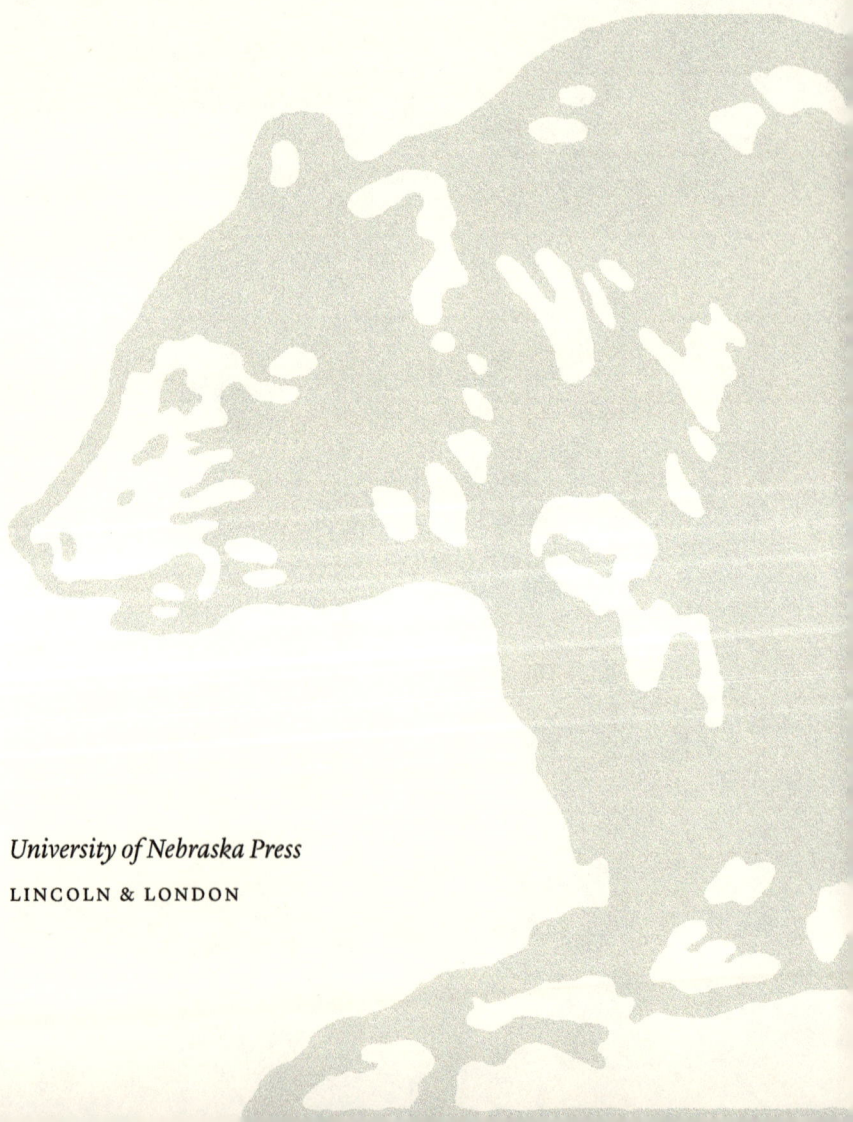

University of Nebraska Press

LINCOLN & LONDON

MICHAEL J. DAX

GRIZZLY WEST

A Failed Attempt to Reintroduce
Grizzly Bears in the Mountain West

Publication of this volume was assisted by a grant
from the Friends of the University of Nebraska
Press.

Library of Congress Cataloging-in-Publication Data

Dax, Michael J.
Grizzly West: a failed attempt to reintroduce grizzly
bears in the mountain West / Michael J. Dax.
pages   cm
Includes bibliographical references and index.
ISBN 978-0-8032-6673-5 (cloth: alk. paper)
ISBN 978-0-8032-7854-7 (epub)
ISBN 978-0-8032-7855-4 (mobi)
ISBN 978-0-8032-7856-1 (pdf)
1. Grizzly bear—Reintroduction—Selway-
Bitterroot Wilderness (Idaho and Mont.)
2. Wildlife conservation—Political aspects—
Selway-Bitterroot Wilderness (Idaho and Mont.)
I. Title.
QL737.C27D337 2015
599.784—dc23
2015005401

Set in Lyon Text by M. Scheer.

Contents

Illustrations

Maps

Acknowledgments

This project began as a term paper during my master's thesis under Dan Flores at the University of Montana. In a seminar class, Dan suggested that someone might write about a little-known attempt to reintroduce grizzly bears to the Selway-Bitterroot Wilderness, which was nearly visible from our classroom. After conducting some preliminary research at Dan's urging, I realized the subject's potential given the complexity of the issue and the multitude of ethical and cultural questions it raised. So first and foremost, I owe an immense debt of gratitude to Dan Flores. His supportive advice and guidance throughout the project was heartening, and without his encouragement, I might not have pursued this project past my master's thesis. Our lengthy conversations about the West and its internal struggles helped me refine my thinking on the subject and develop the narrative arc of the story. Not only was he fully invested in the project from the first term paper through to the final manuscript, but without him, I most likely would have never heard of this forgotten piece of history. Also, when it comes to writing, Dan is a true original whose style is all his own. He encouraged his students to write with a similar freedom, and I hope his influence comes through in my writing.

After piecing a term paper together, I contacted Hank Fischer, a former representative of Defenders of Wildlife who was one of the chief architects of the plan to return grizzlies to the Bitterroots. Hank still lived in Missoula and was kind enough to grant me an interview, during which he offered to give me the box of materials he had collected over the course of the project. The box was sitting in his basement, and his wife had been urging him to get rid of it, so I was happy to take it off his hands. In this box I discovered a haphazard collection of memos, faxes, letters, emails, newspaper clippings, and various publications that no one had laid eyes on in more than a decade. It was beautifully disorganized, and it was all mine. If not for his generous donation, this project very likely would not have become what it did. During my other research efforts, I was able to find a handful of the sources contained within Hank's collection, but no other archive was nearly as robust as the one collecting dust in his basement. Hank remained engaged in my project while I wrote my thesis, and his support and comments were greatly appreciated.

While completing my master's thesis, Jeff Wiltse consistently asked me pointed questions and offered insightful suggestions that made the project stronger. Initially I was hesitant to tackle some of these questions on their merits because they challenged my own beliefs and understanding of the events about which I was writing, and I was nervous for the answers I would find. His questions cut to the core of the dilemma that introducing a potentially dangerous predator to the region created, and they forced me to grapple with difficult issues. I did my best to address his concerns, and the manuscript is undoubtedly stronger because of it. I would also like to thank Michael Patterson who brought a dose of hard science, policy, and firsthand recollection of the issue to my thesis committee. His influence on this book has been subtle, yet important.

Over the course of the project I interviewed a number of people who had been involved with the issue in varying capacities. Thank you to Chris Servheen, Mike Bader, Minette Glaser, Claire Kelly, and Hank Fischer for taking the time to recall events that occurred nearly twenty years ago and to share those memories with me. They filled holes in my narrative that no document alone could possibly have done.

I owe a thank-you to Dylan Huisken and Tom France for reading late drafts of the manuscript and helping me fine-tune my language and analysis. Thanks also to Zack Porter who was kind enough to take photos for the book and provide guidance on some of its other small elements. Barrett Hedges kindly loaned one of his many brilliant photographs for the book as well. I would also like to thank Clinton Lawson for our many conversations that helped me articulate a number of the ideas and theories that appear in the following pages, and for reassuring me throughout this process.

The Matthew Hanson Endowment at the University of Montana awarded me a generous grant that allowed me to continue my research and expand my project from a thesis into a book. This award afforded me the luxury of spending a magnificent summer researching, writing, and occasionally sneaking away to the Bitterroots for solace and restoration.

In this vein I cannot leave out Treasure State Donuts in Missoula, where I spent hours drinking coffee and working on this project the summer after finishing my master's thesis. By giving me free rein over as much counter space as I could possibly occupy, plenty of coffee and donuts, and the ability to continue working after they had closed for the day, I had access to a surprisingly conducive atmosphere for productive writing and creative thought.

I would also like to thank Bridget Barry and the University of Nebraska Press for taking a chance on a first-time author without the typical credentials of someone submitting a manuscript to an academic press. Your confidence gave me the last push I needed to complete the manuscript.

Finally, I would like to acknowledge the importance that the landscape of western Montana and Missoula has had in shaping and inspiring this book. It is close to perfection, and no matter one's position on grizzly bears, everyone who knows it would agree that it is worth fighting for. I will always feel a special connection with western Montana, and the process of writing this book has strengthened that bond. For this I am eternally grateful.

Abbreviations

AWR	Alliance for the Wild Rockies
CAG	Concerned About Grizzlies
CBA	conservation biology alternative
EIS	environmental impact statement
ESA	Endangered Species Act
GBRP	Grizzly Bear Recovery Plan
GYE	Greater Yellowstone ecosystem
IDFG	Idaho Department of Fish and Game
IFIA	Intermountain Forest Industry Association
IGBC	Interagency Grizzly Bear Committee
NCDE	Northern Continental Divide ecosystem
NEPA	National Environmental Policy Act
NREPA	Northern Rockies Ecosystem Protection Act
NWF	National Wildlife Federation
ROOTS	Resource Organization on Timber Supply
USFWS	U.S. Fish and Wildlife Service

GRIZZLY WEST

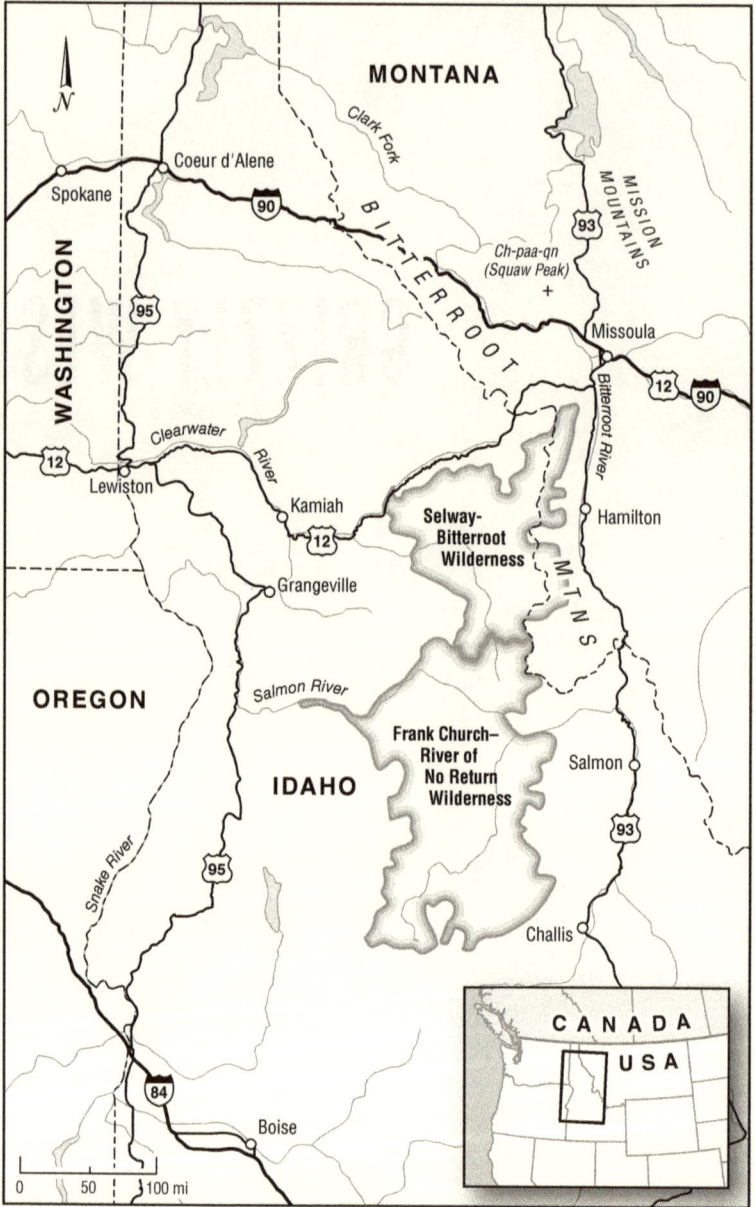

Map 1. Map of Greater Bitterroot ecosystem. Erin Greb Cartography.

Introduction

On a midsummer's morning in 2012, I awoke to a sunny but cool weekend day. The weather was perfect for a hike, so I pointed my car west out of Missoula onto Interstate 90. After a short distance, I exited at the Ninemile Ranger Station, a historic Forest Service outpost that has been preserved in its original state from the 1930s and is open for guided tours. I passed by the row of neat white bungalows, and for the next forty-five minutes bumped my Subaru Outback slowly up the steep, rutted, single-track Forest Service road three thousand vertical feet upward. The U.S. Forest Service built the road nearly a hundred years earlier so logging trucks could transport cut trees to lumber mills, but today the road sees far more sporty, all-terrain vehicles carrying weekend recreationists than it does logging trucks.

Eventually, as I approached the top of the ridgeline, the discreet pullout mentioned in my guidebook appeared, and I parked my car. After the dust from the road settled, I strapped on the new hiking boots I had bought a few days earlier at REI and took off into the beckoning forest. My hike began under a thick, lush cover. Mountain heather blanketed the ground, and I dodged remnant patches of snow along the trail. After a few miles, the trees gave way to a series of steep boulder fields, which were underlined by clumps of bear grass and divided by intermittent stands of whitebark pine trees. The trail eventually directed me up one of the boulder fields, and after a short scramble, I arrived atop Cha-paa-qn. Formerly known as Squaw Peak, the isolated pinnacle dominates Missoula's western skyline and provides a stunning 360-degree view. A few years earlier Montana's state legislature had changed the name in honor of the area's Native American heritage. The new name is from the Salish language, spoken by the local Flathead people, and means "treeless, or shining, peak." No matter its official title, however, the picturesque peak allowed me to see for miles in every direction.

The sky was almost perfectly blue with only a slight breeze to dampen my appreciation for my surroundings. The Forest Service trail crew I had met at the base of the final scramble told me the top of the mountain was less windy than usual, so I counted myself lucky. I ate my lunch, closed my eyes, and basked in the high summer sun before consulting my map to get my bearings. To the north, I could see the Mission Mountains rise from their namesake valley a mere twenty-five miles away, as well as Flathead Lake, the largest natural lake in the Rockies. As I turned south, Missoula's sprawl briefly came into view before a mass of jumbled peaks distracted me and demanded my attention. From my vantage, the Bitterroot Mountains, rising to the west, appeared to stretch for an eternity, beyond the most distant horizon, as far as my eyes could see. They appeared magnificently wild and seemingly untouched by civilization's advances. My mind quickly drifted, imagining the secrets its forested ridgelines and deep valleys knew and kept hidden from distant outsiders. Much of what I saw was part of the Selway-Bitterroot Wilderness, a 1.3-million-acre region established by the Wilderness Act of 1964 that stretches across central Idaho and the far western reaches of Montana. Directly to the south of the Selway-Bitterroot Wilderness is the even larger Frank Church–River of No Return Wilderness. Together, these two wilderness areas make up the largest roadless area in the continental United States and would stir the imagination of even the most stubborn urbanite.

As I stood, almost equidistant between the two mountain chains, my thoughts turned to grizzly bears, creatures that are able to strike fear into the heart of any traveler while simultaneously leaving them awestruck. Despite the fact that the Bitterroots dwarf the Missions in sheer size and extent, I knew that the Missions were home to a healthy population of grizzly bears, and the Bitterroots were not. Why was this the case? This question, I knew, had at least three answers. The first was simple. Since white settlers first started inhabiting the Bitterroot Valley in the 1850s, they had pursued the destruction of the region's grizzly bears with unparalleled zeal until the middle of the twentieth century, when a Forest Service ranger recorded the last confirmed sighting. The region surrounding the Mission Mountains saw far less white settlement, and the Confederated Salish and Kootenai Tribes, whose reservation borders the western side of the Missions, coexisted relatively peacefully with the bears. The second answer was slightly more complicated. The Mission Mountains were isolated enough that grizzlies had been able to persist there, but the advances of modern infrastructure had proven significant enough to prevent them from migrating south into the Bitterroots. Fewer than fifty miles and only two major roads separated the two mountain chains, but no grizzly had yet made the trek.[1] The third answer was the most complex—and is the subject of this book

In accordance with the Endangered Species Act (ESA), the U.S. Fish and Wildlife Service (USFWS), along with a number of nongovernment organizations and advocacy groups, had formulated a plan in the mid-1990s to bring grizzly bears back to the Selway-Bitterroot Wilderness. Efforts began just as the long, hard battle for gray wolf reintroduction in Yellowstone National Park and central Idaho was nearing success, and many of the people who developed the plan for Bitterroot grizzlies had worked tirelessly on wolf recovery throughout the 1980s and 1990s. Wolf reintroduction had been extremely divisive as western politicians and ranchers, who feared that wolves would prey on their livestock, fought the action for nearly twenty years. Relying on heavy-handed tactics developed by the environmental movement during the 1960s and 1970s, wolf advocates eventually prevailed, but for a few—including Hank Fischer of Defenders of Wildlife and Tom

France of the National Wildlife Federation (NWF)—the long, drawn-out process had thoroughly discouraged them, and they wanted to find a way for grizzly reintroduction to avoid the intense, intransigent debates that had plagued wolf recovery.[2]

To this end, Fischer and France helped develop a plan for grizzly bears that advocated reintroducing them to the region under an innovative set of guidelines that not only relaxed the standards of the ESA but also allowed for local control on an unprecedented scale. Instead of classifying the reintroduced bears as fully "threatened," the plan chose to consider them as "experimental, non-essential," an idea that had been critical to the successful implementation of wolf recovery. Additionally, whereas the federal government had managed every previous endangered species program—including wolves—the plan to reintroduce grizzlies to the Bitterroots proposed putting this authority into the hands of the states and citizens through the creation of a citizen management committee. Over the course of the previous decades, conservatives in the West and throughout the country had complained of too much federal oversight from what they considered a distant bureaucracy that neither understood nor respected their way of life.[3] The new plan was a direct response to those criticisms, and its architects hoped this novel approach would mollify the plan's attackers and win over critics.

Furthermore, unlike with wolf reintroduction and other controversial endangered species recovery programs, such as the spotted owl of the Pacific Northwest, environmentalists now worked hand in hand with Idaho's timber industry to develop the plan for grizzlies. The timber industry feared that if grizzly bears came back to the Bitterroots as a fully "threatened" species, environmentalists would be able to use the ESA to keep loggers out of the forests in order to protect the bears' habitat. With this in mind, two groups that represented Idaho's timber industry, Resource Organization on Timber Supply (ROOTS) and the Intermountain Forest Industry Association (IFIA), broke with tradition. Instead of opposing reintroduction at all costs only to face the possibility of losing in the end, they approached the problem proactively. Working with Fischer and France, they formulated what came to be known as the ROOTS plan, or citizen management alternative, which would bring grizzlies back to the Bitterroots while preserving the timber industry's presence in the region.[4]

With this unlikely coalition in place, the plan earned praise from both regional and national observers, and its advocates believed their citizen management model would inspire future endangered species restorations across the country. As a result, the plan progressed through bureaucratic channels much quicker than had wolf recovery before it. Within four years of the USFWS's decision to pursue grizzly recovery in the Bitterroots, Congress allotted funding to prepare an environmental impact statement, the process required by the National Environmental Policy Act (NEPA) to ensure public input when any significant changes are recommended to be made to an ecosystem. Wolf recovery had needed sixteen years to get that far, and without the innovative approach taken by the ROOTS coalition, grizzly reintroduction most likely would have experienced similar delays.[5]

The plan, however, was not without its critics. Surprisingly, some of the plan's staunchest opponents were on the left. These environmentalists were loyal to the model of reform set forth by the environmental movement of the 1960s and 1970s and were opposed to the ROOTS plan because they believed it made too many concessions and was biologically risky. They had developed their tactics in an era in which environmentalism had a mandate from the American public that allowed it to press forward boldly, without significant political resistance. These groups—which included the Sierra Club, Audubon Society, and a regional organization called the Alliance for the Wild Rockies, along with others—championed their own plan, which they called the conservation biology alternative (CBA), and were unwilling to make any meaningful biological compromises for the sake of politics. Their plan advocated natural migration instead of reintroduction, allowed the bears to maintain their fully threatened status, and mandated extensive habitat restoration to ensure the grizzlies' long-term survival. Although the plan gained substantial support in urban centers such as Missoula, Boise, and Helena, it gained little traction in the rural areas of the region where people still worked in extractive industries. Nevertheless, this bloc provided substantial resistance that created hesitancy within the USFWS to support the ROOTS plan and contributed to the plan's eventual failure.[6]

The ROOTS plan encountered additional opposition from conservatives who did not want to see grizzly bears return to the Bitterroots under any circumstances. These critics had watched as the environmental movement gained more and more power over the preceding decades, culminating with the reintroduction of wolves, and they were determined to push back against its progress. For them, grizzly bears posed an imminent safety concern, and they did not trust the safeguards the architects of the ROOTS plan had put in place. Similar to the proponents of the conservation biology alternative, conservative critics were unwilling to compromise in any way, and their opposition was most directly responsible for the plan's collapse.

These opponents were part of a group scholars have identified as the Old West, referring to the culture and economy that took shape when white settlers moved into the region in the mid- to late nineteenth century. Most of these settlers engaged in and were reliant upon extractive industries. Logging, ranching, farming, and mining became the backbone of the West's economy as the region became a natural resource colony for much of the eastern half of the country. Because the livelihood of Old Westerners relied so heavily on extractive industries, their relationship with the land justified its protection only so far as was necessary to exploit it for economic gain in a sustainable manner. This anthropocentric ethic, known as conservation, first gained prominence at the turn of the twentieth century—due in part to President Theodore Roosevelt and his chief forester, Gifford Pinchot—and directed the country's natural resource policies over the first half of the century. While Old Westerners eventually embraced conservation, they continued to view the federal government, which maintained ownership of a substantial portion of land in the West, as the overbearing parent they could not escape. As a result, many Old Westerners developed an antiauthoritarian mentality and were inherently suspicious of any action on the part of the federal government.[7]

Many of the environmentalists who supported the conservation biology alternative identified with the New West, which began establishing itself during the 1960s and 1970s when tourism boomed in the post–World War II era and environmentalism replaced conservation as the dominant land ethic. As a result of hitting the road in the postwar years, Americans visiting the West's scenic wonders began to develop a new environmental ethic that went beyond conservation, and they began migrating to and inhabiting the region from which they drew this inspiration. This new population desired to protect the environment not only for the sake of harvesting its resources but also because of the intangible, spiritual benefits it afforded in its natural state. This wave of settlers was also a part of the country's new upper class. Beginning in the late 1950s and 1960s, access to upper-class status had shifted away from its hereditary roots and become increasingly dependent on merit and education, and many of these new elites supported the environmental movement. When they moved west to settle on their personal slices of utopia, the new elite brought this changing set of values with them. As a result, the region experienced a wholesale social and cultural shift with implications that went beyond environmental ethics.[8]

In reaction to these changes, Old Westerners, especially in the Northern Rockies, pushed back against this new population and its values. Although environmentalism had taken the reins of federal policies by the 1970s, Old West conservationists had not converted to this new ideology. But the environmental movement enjoyed such widespread popularity that through the first half of the decade Old Westerners had little chance of making their complaints heard. By the end of the 1970s, however, the country had grown disillusioned with the federal government, the nation's economy had declined, and the wealth and abundance that had made the environmental movement possible over the previous fifteen years gave way to stagflation and a

subsequent conservative backlash. In 1979 a faction of Old Westerners took advantage of this shift in the political climate and launched
what many termed the Sagebrush Rebellion. The rebellion was an
attempt by a number of western states to wrest control of public lands
away from the federal government and manage them in accordance
with the principles of nineteenth-century conservation. The movement thrived for a few years but collapsed by the mid-1980s. Nonetheless, sentiments from the movement persisted and its adherents
reorganized themselves by the end of the decade and into the 1990s
under a variety of different names, including the Wise Use Movement,
Property Rights Movement, and County Rights Movement. These
factions represented Old West extractive industries that wished to
take control away from the New West and its environmental policies. The debate over grizzly reintroduction by no means created
this conflict but rather intensified it and brought it out in the open.[9]

In addition to wanting to ensure their economic interests would
remain a priority in public policy discussions, Old Westerners disdained the rise of the New West because they viewed its economy
and conceptions of land use as inauthentic, and they believed the
ascension of the New West would mean the end of the region's exceptionalism. In the American mind, the West had long been a special
place with which people identified, and Old Westerners believed the
region's identity would lose its distinctiveness if the New West continued to flourish. The divide between the two groups cut to the core of
what it meant to be a westerner, and during every controversy that pit
the two sides against each other, ultimate control of the region seemed
to be at stake.[10] While many people loyal to the Old West had a number of specific critiques of the ROOTS plan—including charges that
the citizen management committee would not have any real power,
fears that grizzly bears would threaten personal safety, and accusations that the project would cost too much—their criticisms essentially
boiled down to anxiety over the possibility that their hundred-year
reign over the region would end, and the West's unique character as
they knew it would wither.

Despite intense opposition from both the conservative Old West and the liberal New West, the ROOTS plan marched slowly forward over the closing years of the twentieth century. The Endangered Species Act required the federal government to recover grizzly bears in the Bitterroot ecosystem, so despite local opposition to the proposal, the USFWS was able to move forward with the plan. Throughout the compilation of the environmental impact statement, the ROOTS coalition stressed the need to streamline the process while the Clinton administration, which was sympathetic to environmental reform, was still in office, but delays created by hard-lined New West environmentalists squandered this opportunity. Not until November 2000, after concluding the six-year environmental impact statement process, did the USFWS finally approve the proposal and plan to start relocating bears to the Bitterroots in the summer of 2002.

When the controversial presidential election of 2000 reached the Supreme Court, bear advocates nervously awaited the court's ruling, knowing that its decision would effectively decide the fate of their project. If Democratic candidate Al Gore—whose sympathies for environmentalism aligned with those of the New West—won, funding grizzly bears would continue to be a spending priority. But if Republican candidate George W. Bush, who identified strongly with the Old West, became president, the project was as good as dead. Soon after Bush took office, in January 2001, the new secretary of the interior heeded the wishes of Idaho's Republican governor, Dirk Kempthorne—who opposed "bringing these massive, flesh-eating carnivores into Idaho"—and put the project on hold indefinitely.[11] Observers at the time were quick to explain the project's collapse as a result of partisan politics, but in retrospect, the ultimate source of the plan's failure was embedded much deeper in the dichotomous nature of the divide between the Old West and the New West.

The debate over grizzly bears revealed and highlighted all the ways in which the Old West and New West clashed over conceptions of land use and the role of humans in the natural world. While partisan politics suffices as a superficial explanation for the project's demise, the reason Governor Kempthorne so fiercely opposed it and President Bush, who had previously been governor of Texas, was so ready to accommodate his wishes, originated with this divide. Pioneers who first settled the region considered it their duty to subdue the land, rid it of any wild character, and assert their supremacy. The conservation ethic that emerged at the turn of the twentieth century encouraged Old Westerners to protect the land's resources in a sustainable manner, but it did not fundamentally challenge their prior relationship with the land. Rather, its rationale was purely utilitarian, and conservationists protected the land only to the extent that was necessary to make a sustainable living. Therefore, the conservation ethic made no room for grizzly bears and actually necessitated their destruction. Grizzly bears not only put human safety at risk but they also threatened ranchers' economic livelihoods by preying on livestock. As a result, Old Westerners, with help from the federal government, exterminated the region's grizzly bear population over the course of the late nineteenth and early twentieth centuries, leaving grizzly bears to exist on just 2 percent of their former range.[12]

As the New West took shape in the second half of the twentieth century, and environmentalism began to replace conservationism as the nation's dominant environmental ethic, attitudes toward grizzly bears changed. The rationale employed to protect the land and its resources transitioned from purely economic concerns to include romantic, spiritual, and ecological values. New Westerners and environmentalists across the country sought to protect animals like grizzly bears because they were instrumental to maintaining ecological balance and because they represented the wildness that these groups considered authentic and necessary for preserving the American spirit. Also, as ecotourism gained an increasing hold over the region's economy, the opportunity to see animals like grizzly bears began attracting visitors from across the globe. Whereas grizzly bears had been a hindrance to the economy of the Old West, they became an asset to the growing economy of the New West.[13]

For Old Westerners, who by the 1980s were predominately conservative Republicans, reintroducing grizzlies was not only a rejection of their values, it suggested they may have been wrong to extermi- nate them in the first place. Furthermore, such a drastic shift in policy from killing bears to reintroducing them had potential implications for their continued political influence on a broader scale. As a result, reintroducing grizzlies was not just about bears, it was about power, and keeping grizzlies out of the region went hand in hand with safe- guarding their hegemony. For New Westerners, who mostly aligned with the Democratic Party, reintroducing grizzly bears was no less about power. In addition to the high-minded rhetoric that consid- ered reintroducing grizzly bears as "an exercise in humility" that would give us "some hope for the earth," reintroduction also repre- sented an urge on the part of the region's new and rapidly expand- ing demographic to become the West's dominant interest group.[14] So while this account will first and foremost be about grizzly bears, it will also be about power and the changing political, economic, and cultural dynamics of the West at the turn of the twenty-first century.

Although the New West's economy had surpassed the Old West's extractive economy by the year 2000, the ideology and political agenda of the Old West retained far more power than its share of the economy may have suggested. Following the release of wolves in Yellowstone and central Idaho, New Westerners celebrated. After a long, hard battle, Old Westerners had relented, unable to prevent the wolves' return any longer, and New Westerners believed their victory represented a larger triumph against the Old West. Their ideology had prevailed, and from that point forward, they would have the upper hand in regional politics.[15] Any New Westerners who believed this, however, vastly overestimated their victory. While the New West's political potency had certainly grown, the Old West still held a sig- nificant grip over the region's politics. And even though Old Western- ers had been unable to prevent wolf reintroduction from proceeding, they were nowhere close to allowing their control over the region to slip from their grip permanently. As debate over Bitterroot griz- zly reintroduction heated up, their sense of urgency mushroomed.

They refused to relent, determined to force the region to recognize the power of their interests.

One of the reasons why the Old West was able to maintain its political power lay in the fact that so much of the region's identity was wrapped up in a sense of nostalgia and history that was based on the region's frontier persona. Mountain men, trappers, cowboys, and loggers were still the images that most Americans associated with the West. Even New Westerners appropriated many of the symbols and styles that originated in the Old West, such as cowboy boots and fringed jackets. As a result, the region was defined by and wedded to its traditions, which in turn trickled down to its political culture. Claims to this "authentic" identity were a source of power, and Old Westerners were incapable of abandoning the philosophies that had been guiding their actions over the previous century. For Old Westerners, extractive industries and the culture they nurtured were authentically western, and many believed the region's exceptionalism stemmed from them.[16] Old Westerners were convinced that maintaining a connection to this heritage would be paramount to preserving this exceptionalism. Claims to Old West identity, therefore, became a source of capital, necessary for wielding power, especially political power, and Governor Kempthorne's reaction exemplified this need western politicians felt to gravitate toward the Old West's interests.

Tensions between the Old West and the New West run through many stories dealing with the modern West, but the failure of Bitterroot grizzly reintroduction reflects another facet of this dilemma that is true of American politics at large. The timber industry representatives who helped develop the ROOTS plan were part of the Old West, just as the representatives from Defenders of Wildlife and the National Wildlife Federation represented the New West, but together, these groups stepped away from their ideological bases to find a solution that united, rather than divided, their respective sides. For a short while, this strategy was effective, but in the end, it could not unite these two disparate foes. The region was locked in a dichotomous battle that had no room for a third party. Just as Old Westerners felt a need to dig in their heels following their defeat over wolf recovery, many New Westerners wanted to capitalize on the momentum created by this victory and became just as determined as their Old West counterparts not to give an inch when it came to Bitterroot grizzlies. As a result, they were unwilling to accept the compromises of the ROOTS plan, which created significant delays in the environmental impact statement process that ultimately proved detrimental. While intransigence on the part of Old Westerners was the most direct cause of the proposal's collapse, the project ultimately failed because of the unyielding nature of the *divide* between the two groups, not because of one side in particular.

The increasing intractability of this divide was a regional issue that reflected larger themes of national politics and culture. Finding room for compromise between the ideals of the Old West and the New was nearly insurmountable in some cases, but as politics in the nation grew increasingly contentious throughout the 1990s, forging diverse coalitions became ever more difficult. Even today, bipartisanship remains an ostensible goal and a watchword among politicians, but genuine bipartisan cooperation is almost unheard of as few Democrats or Republicans are willing to step across the aisle.[17] During the 1990s, the environmental movement had begun to fracture as it lost the cachet that had propelled it during the 1960s when it could easily cast itself as David facing off against Goliath. Some environmental organizations adapted to the changing times and sought solutions based upon compromise, while other groups renewed their commitment to the same goals and strategies forged during the environmental movement of the sixties and seventies. Pragmatism and collaboration had recently emerged as new strategies in natural resource management and would become increasingly common over succeeding decades, but at the time, grizzly bears proved too divisive an issue to rise above the tumultuous political climate that characterized the West at the verge of the twenty-first century. While this account will be first and foremost about the grizzly bear reintroduction that almost was, it will also be about the failure of compromise in a political system that prides itself on its ability to cooperate and find common-sense solutions.

Throughout the 1990s, news reports concerning grizzly reintroduction proliferated as almost everyone throughout Idaho and Montana seemed to have an opinion on grizzlies. Rarely did a few weeks pass without an editorial or news article appearing in one of the region's papers. News outlets throughout the country regularly reported on the issue as well. Newspapers in Arizona, Oregon, Alaska, New York, and Washington DC all published stories on the prospective reintroduction, and ABC News even aired an extended broadcast on the issue. Twelve years after the plan suddenly crashed and burned, however, few people in Montana, Idaho, or elsewhere across the country seem to remember the event. Wolf reintroduction remains in the news, fully capable of sparking controversy anytime someone expresses an opinion; but a story about how the U.S. Fish and Wildlife Service had actually approved a plan to reintroduce grizzly bears into the Bitterroots at best garners a passive curiosity. Similarly, while writers have waxed at length about the events and significance of gray wolf reintroduction, no one has yet written any comprehensive account of the Bitterroot grizzly plan. In his 2008 book, *Grizzly Wars: The Public Fight over the Great Bear*, David Knibb devotes a few pages to the details of the Bitterroot grizzly reintroduction, but he makes few significant contributions to the discourse beyond how it affected grizzly recovery in Washington's North Cascades, which was the main thrust of his book. As a result, *Grizzly West* will be the first historical account of what was nearly the most significant endangered species story of the 1990s.[18]

Other writers have expounded at length about the rise of the New West, and although many are critical of some of its contradictions—cheaply commodifying and glorifying a culture that it is replacing—they do not sufficiently examine the ongoing tension between it and the Old West in regional politics. In her essay, "Way Out West . . .

Ghost Towns, Gray Wolves, Territorial Prisons & More!," in *Imagining the Big Open*, Karen Jones discusses the transformation of the wolf in the popular mind, but she minimizes the influence of the Old West's continued resistance to wolf recovery. Similarly, in his introductory essay "Paradise Revised" for *Atlas of the New West*, Charles Wilkinson argues that there could still be a place for extractive industries in the region; but he does not seem to consider how drastically the polemical nature of regional politics has affected the West. He implies that communities could easily and simply choose which model to follow.[19] My account of the debate over grizzly reintroduction will attempt to correct these various oversights and argue for an Old West that, through the 1990s and into the 2000s, retained much of its political power even if its cultural and economic presence had waned.

Finally, a note on terms. Today, or since the 1990s, few wildlife advocates refer to themselves as environmentalists. Rather, they generally refer to themselves as conservationists. But in an effort to distinguish between people who adhere to an early twentieth-century ethic of conservation and those who are more loyal to 1960s and 1970s environmentalism, I refer to the latter as environmentalists. Also throughout the book, I use the word "ethic" to mean any system of beliefs used to determine right from wrong. This system could be based on economics, altruism, power, morals, or any combination of these factors. However, I do not mean for ethics to be synonymous with morals. Next, the terms Old West and New West are imperfect. While the New West is characterized largely by those who have moved to the West in their own lifetimes to enjoy the region's natural amenities, someone whose family has lived in the West for multiple generations may still identify with the New West. At the same time, the values of someone who moved to the region in his or her own lifetime may better align with the Old West. Generally, this is not the case, and the terms are more than suitable for how I will use them, but this is an important caveat. Finally, throughout the book, I occasionally reference black bears as well as grizzlies, and many times, I simply refer to "bears." Unless using the term "black bear" specifically, the common reference to "bears" should be taken to mean grizzlies only, and not all bears.

After about a half hour on top of Cha-paa-qn, contemplating how the two mountain ranges reflect dichotomous visions of the West, I headed back down the mountain, retracing my steps to my car. The drive back to Missoula seemed shorter, and once back in town, I headed to my favorite local microbrewery. After a few pints of Missoula's favorite beer, Cold Smoke Scotch Ale ("cold smoke" being the highest quality of powder snow lusted after by ski bums), I headed out across town to an annual Celtic music festival. As I crossed the bridge that spans the Clark Fork, a number of kayakers weaved in and out of Brennan's Wave, a rapid built by the city ten years earlier to accommodate local recreationists. On the other side of the bridge, farther upstream, a fly-fisherman casted out into the river, hoping to hook a trout before darkness settled in. There it was, in front of my eyes: the New West. The Old West was still fighting for the right to assert control over the region, and many politicians continued to pay homage to this heritage, but culturally and economically, the West had undoubtedly changed. How these dramatic shifts would influence the future presence of grizzly bears in the Bitterroots and elsewhere was still uncertain.

1

Grizzly Americana

After tracking the bear for two days, William Wright finally spotted it a few hundred yards above the edge of a canyon, deep in the Bitterroot Mountains. He was disappointed at the bear's diminutive size, but decided to kill it anyway. Wright positioned himself downhill from the bear so he would have the best possible shot. When he eventually reached the edge of the canyon, he looked into it and saw another bear—the biggest bear he had ever seen. This was the bear he had been tracking. Wright had to think for only a moment before deciding to kill them both. He repositioned himself, took aim at the larger of the two bears and pulled the trigger. Wright then turned to the other bear, which had stood erect to investigate the noise. He quickly aimed and pulled the trigger again. This bear, like the first, dropped without taking another step. As Wright began walking over to admire his kills, he heard rustling in the next ravine. When he looked up, he discovered the source of the racket was a sow grizzly and her two cubs. Once again, Wright instinctively dropped to his knee, reloaded, took aim and fired. He instantly downed the sow and only needed two more shots to kill the bewildered cubs. He had killed five bears with five shots in five minutes. According to Wright, "This was the greatest bag of grizzlies that I ever made single-handed."[1]

Originally from New Hampshire, William H. Wright moved west at an early age with a desire for adventure. Ever since he was a child, grizzly bears had fascinated him, so when he moved to Washington, Wright dreamed of killing one of the great beasts for himself. He relocated to Spokane in 1883, and over the next six years he made some preliminary attempts at hunting grizzlies. However, when he moved to Missoula, Montana, in 1889, he had still never seen one. Nevertheless, his desire had not diminished, and he headed southwest into the Bitterroots the next summer with a string of packhorses and enough supplies to last the season. Although success did not come easily, by summer's end Wright had shot his first grizzly. Over the next few years, his hunting prowess improved immensely, and he became one of the region's most prolific grizzly hunters. In addition to his amazing account of five in five shots, Wright achieved other astounding feats of grizzly bear hunting. On a single trip in the Bitterroots in 1895 he shot thirteen grizzlies, and on an earlier hunt, he and a few friends killed four bears in a single evening. Despite these achievements, by 1897, Wright's curiosity and appreciation for grizzlies had overcome his desire to hunt them, and he exchanged his gun for a camera. Wright dedicated the remainder of his life to studying and protecting the bear, but he was the exception rather than the rule, and by the time he died in 1934, the Bitterroots were all but devoid of the great bear.[2]

Despite grizzly bears' presence in the mountains and meadows of the Bitterroots since time immemorial, the region's grizzly population was practically nonexistent within four decades of Wright's arrival. Over that period, sport hunters like Wright, ranchers, and other settlers systematically eliminated the region's grizzly population. Although black bears proliferated in the area throughout the twentieth century, the last confirmed grizzly sighting in the Bitterroots had long passed by the mid-twentieth century. Such was the case across the Rocky Mountain West where pioneers sought to tame the land in the name of progress. As settlers poured into the region over the nineteenth and twentieth centuries, they transformed its more wild characteristics to make the region suitable for farming, ranching, logging, and mining, which meant pushing out much of

the native wildlife. Although almost all species suffered, the potential threat to human supremacy that grizzly bears represented earned
them a distinct reputation and distinguished them from other reviled
species such as wolves, coyotes, and mountain lions. By the eve of
World War II, with the help of the government and its wise-use conservation policies, the West had methodically and unapologetically
reduced or exterminated all of these major predators.

Such was the case across the West and the country at large. Over
the nineteenth century, Euro-Americans doggedly pursued taming
the wild character of the region, asserting their dominance over the
landscape. Killing grizzly bears was a small but essential piece of a
much larger campaign to ensure the nation's supremacy over man
and nature. By the end of the century, conservationists like chief forester Gifford Pinchot and President Theodore Roosevelt introduced
the principles of conservation and wise use to the country, which had
been extracting its natural resources at a hell-fire pace. Wise use—
the idea that natural resources should be managed sustainably so
that future generations would have similar opportunities to benefit
from the land's abundance—became the dominant ethic governing
the nation's natural resources. However, this utilitarian philosophy
justified actions in terms of economic and tangible benefits and left
the reigning ideology largely intact. Predators like grizzlies threatened both people and their livelihoods, and the federal government
actively pursued their destruction, forcing the country's grizzly population toward the brink of extinction.

Through the first half of the twentieth century, the persecution
of grizzly bears continued unabated. However, in the postwar era,
grizzly bears' status in the American mind began to improve. As
more Americans traveled westward, visiting national parks, grizzlies
became major attractions for eastern and European tourists. Combined with the emerging environmental movement, many Americans
quit viewing the bear as a burden to progress and economic stability that needed to be eradicated, regarding it instead as a national
treasure that deserved to be protected. Additionally, as pop culture

increasingly appropriated the bear for its purposes, grizzlies were recast as friendly, not ferocious. By the time Congress passed the Endangered Species Act in 1973, grizzly bears had become an iconic symbol of the American West, earning themselves protection under the new law and cementing the vast transformation in image and treatment they had achieved over the previous century. In this way, grizzlies provide an ideal lens through which to examine and understand the significance and scope of the nation's transition from conservation to environmentalism.

Like so many pieces of western history, Euro-Americans' relationship with the grizzly can be traced back to Lewis and Clark's Corps of Discovery. When the group saw its first grizzly bear on October 20, 1804, in present-day North Dakota, it made history. No Euro-American had ever before documented a grizzly bear sighting, so for all intents and purposes, it was history's first. Peter Cruzatte, the expedition's hunter, took the first shot at the bear, and after firing, he promptly became the first American to run from a grizzly, forgetting both his hatchet and gun in the process. Cruzatte survived the ordeal, but his close call had little effect in dissuading the party from shooting at more bears along the way. Over the course of the trip, the expedition saw dozens of grizzlies, many of which it killed. Even after a few more close calls in which wounded bears chased after their attackers, the expedition could not help but try to kill as many bears as possible. The enormous bear fascinated them, as it would many Americans over the next two hundred years, but the ideology that drove their interest dictated that they kill as many bears as possible. In these interactions, one can see the conflicting forces of fear and the desire to dominate and subdue nature that would govern Americans' relationship with the bear over the next 150 years. Historian Dan Flores identified in these interactions an "optimistic arrogance, coupled with a typical American faith in scientific technology's ability to prevail."[3] While a modern observer may be left perplexed at the Corps's unyielding thirst for killing bears, it was this exact attitude that epitomized how Americans would interact with wildlife, especially apex carnivores, over the remainder of the nineteenth century.

By the time they reached the Bitterroots, the Corps's desire to kill the great bear had not diminished, and while passing through the Bitterroots, both westward and eastward, they shot multiple grizzlies. On their way to the Pacific Ocean, the explorers stayed at Traveler's Rest, a few miles south of present-day Missoula, at the foot of the Bitterroot Mountains. While camped, the expedition shot two bears of undetermined species that very possibly could have been grizzlies. As they left their campsite, they passed a tree from which the hide of a grizzly hung—most likely placed there by local Indians. On their return trip, the expedition killed six grizzlies in the vicinity of present-day Kamiah, Idaho, on the western edge of today's Clearwater National Forest. Without a doubt, when Lewis and Clark passed through the Bitterroots and over Lolo Pass at the dawn of the nineteenth century, grizzly bears occupied the region in significant numbers. Yet within the next century, the bears' status as lords of the West had been usurped by a population of settlers that was successfully waging a campaign for their total annihilation.[4]

In contrast to the hostile position that Lewis and Clark and later Euro-Americans assumed in relation to the grizzly bear, Native American tribes throughout the West that had lived among the bear for thousands of years had a much different relationship with them. Meriwether Lewis himself commented in his journal that the Shoshone considered killing a grizzly to be a great honor, and they wore the bear's claws around their necks so that its spiritual power would transfer to them. The Navajo of the Southwest never dared attack a grizzly unprovoked, fearing its retribution. However, if a bear killed a Navajo, an armed party would find the den of the guilty bear, and before killing it, would perform an elaborate ceremony asking the bear's forgiveness in advance. Other tribes frequently depicted bears as essentially human and valued them at the individual level, not just as a population. A story from a California tribe tells of a bear playing a friendly game of cat and mouse with a man who was scared for his

life. The short tale assigns the bear a sense of humor not often associated with wild animals and demonstrates the high regard with which
this tribe held the bear. More relevant to how Native populations in
the Bitterroot viewed the bear, a Nez Perce legend suggests that the
famous Lolo Trail, which the Corps of Discovery used to cross the for-
midable mountain range, was established when a grizzly bear guided
a lost boy through those mountains. That the Nez Perce credited a
grizzly bear with this important discovery clearly indicates the high
esteem in which they held the bear. Every tribe that encountered
grizzlies did so in a slightly different manner, and many did, in fact,
kill them, but none of these tribes either questioned the right of the
bear to exist or sought its absolute destruction.[5]

Nevertheless, Indians' relationship with the bear had little impact
on how Americans viewed the great beast. In 1815 George Ord, a
member of the Academy of Natural Sciences and the American Philosophical Society in Philadelphia, using Lewis and Clark's field reports
as his basis, assigned the grizzly its scientific name, *Ursus horribilis*.
While Ord had never seen the bear himself, the species name *horribilis* helped solidify the bear's negative reputation, and future generations of Americans would use this moniker as ammunition to justify
the bear's destruction.[6] While this name was not wholly deserved,
grizzlies, more than any other nonhuman species, challenged Americans' westward expansion and few saw any room for the great bear
in what they envisioned for the region's future.

Much like the indigenous peoples who experienced similar treatment by Euro-Americans, grizzlies originally inhabited the forests
of Asia and began migrating across the Bering Land Bridge roughly
fifty thousand years ago. Somewhere in this process, they adapted
to their new home, which was unlike the forests of Asia, and evolved
from carnivores to omnivores. Today grizzlies range in size depend-
ing on food sources, with bears living in coastal regions being larger
than inland bears because of the availability of fish and richer vegetation. Those in the Northern Rockies typically range from 300 to 450
pounds, with males weighing in at the upper end of that range. In the

Northern Rockies, grizzlies rely heavily on fruits, insects, nuts, and roots for the majority of their diet. They are opportunistic foragers, and while they will prey on elk and deer calves in the spring, they do not hunt for most of their food in a traditional sense.[7]

Even so, as more Americans poured west after Lewis and Clark's return, interactions between humans and bears increased. Sensationalist stories of these incidents proliferated, and the grizzly's reputation as a menacing beast quickly grew into legend. In his 1836 book *Astoria; or Anecdotes of an Enterprise Beyond the Rocky Mountains*, Washington Irving recounted a handful of instances in which his party encountered one of these infamous bears. On one occasion, a hunter named William Cannon was returning to camp with meat from a buffalo he had just shot. While passing through a narrow ravine, Cannon turned to find a grizzly bear in hot pursuit of him and the fresh meat he was toting. Having heard that the bear was invulnerable, Cannon did not even consider trying to shoot it and instead ran for his life. The panicked hunter climbed a tree, where he spent the night, unsure if his assailant lay waiting for him. But when he awoke, the bear was gone and Cannon returned to camp without further incident.[8]

In another account, famed western writer George Frederick Ruxton relayed the story of an episode in which mountain man Hugh Glass found himself at the mercy of a ferocious grizzly. Glass and his partner had come across the bear while traveling through the Black Hills, and once spotted, both men took a shot at the bear. Unfortunately for them, the shots had little effect beyond incensing the grizzly and persuading it to charge the ill-prepared attackers. Both men ran, naturally, but Glass tripped on a rock, and by the time he arose, the bear was towering over him on its hind legs, snarling and baring its teeth. Glass's companion shot at the bear again, but his shot failed to discourage the bear's attack. Glass pulled his knife and as the bear fell on him, Glass stabbed the bear repeatedly while the two wrestled each other, fighting for their lives. Glass's companion ran back to camp to get help, and when the group returned, they found a severely wounded Glass lying next to the dead bear, which had more than twenty lacerations in addition to the three gunshot wounds.[9]

Both of these stories sensationalized the ferocity of the great bear, stressed the danger that all humans faced while in its presence, and helped perpetuate the notion that the only good bear was a dead bear. As settlers continued to move westward and establish permanent homes—ranching, logging, and farming—any tolerance for these threatening animals evaporated. Grizzlies' size and potential lethality threatened human primacy leading settlers to kill grizzly bears out of a practical need for personal safety as well as ideological dominance. Additionally, because grizzlies preyed on livestock and competed with humans for other food sources, the need for their destruction contained an economic component as well. Aided by lurid tales recounted in dime novels, however, these real concerns developed larger-than-life dimensions, and grizzlies quickly gained reputations for being far more rapacious than they actually were. The fact that grizzlies often scavenged for food undermined their reputation even further. Americans considered such behavior dishonorable, so destroying them assumed a moral dimension that invigorated the campaign against them. As the nineteenth century wore on, fear and hatred led the settlers' crusade to transcend practicality and become culturally ingrained to the extent that the destruction of every last bear became crucial. And so, throughout the nineteenth and early twentieth centuries, Americans' relationship with the grizzly accorded with a model that unashamedly required the bear's unconditional annihilation.[10]

Eventually the market economy and government policies evolved to accommodate this cultural development. Grizzly bear hides were never a prized commodity like the pelts of bison or beaver, but because settlers killed them en masse their hides made it to market anyway, fetching as much as four dollars per hide, and local governments found ways of encouraging grizzly hunters. Between 1827 and 1859, Hudson's Bay Company outposts in the Northwest shipped almost 3,800 grizzly hides to eastern markets. A Rocky Mountain trapper named George C. Yount, who first began trapping in 1826, bragged that he "often killed as many as five or six [grizzlies] in one day." Although this campaign hardly needed any encouragement, in 1893, the territories of Arizona and New Mexico passed a law that allowed counties to offer bounties on a number of animals, including grizzly bears. The U.S. Biological Survey, which was the federal agency tasked with eradicating predators in the West for the benefit of ranchers, occasionally targeted bears, but mostly focused on wolves. Just the same, the poison its agents used to lace bait stations intended for wolves killed curious grizzlies just as effectively, and within the first few decades of the twentieth century, their populations had plunged.[11]

Amidst this craze of killing and destruction a few voices spoke out in favor of the bear's protection. John "Grizzly" Adams is perhaps the most famous early friend of the great bear. Adams traveled west as part of the California gold rush, but after a few seasons in the mining camps, he took to the mountains where he captured, tamed, and befriended a number of grizzly bears. Adams was known for playfully wrestling with these bears, and one of his companions, Ben Franklin, even came to his rescue when another grizzly attacked Adams in 1855. At the turn of the nineteenth century, Enos A. Mills, a contemporary of William Wright, gained prominence as a naturalist and wrote many descriptive passages that celebrated the majesty of the grizzly. "Perpetuate the grizzly in our wild places and National Parks, and this will fill all wild scenes again with their appealing primeval spell—the master touch which stirs the imagination. . . . The imagination will be alive so long as the grizzly lives," wrote Mills at the close of his 1919 book, *The Grizzly: Our Greatest Wild Animal*. Despite romantic passages such as this, pro-grizzly sentiments were in the minority, and their populations suffered as a result.[12]

More common was Theodore Roosevelt's account of shooting his first grizzly. Roosevelt appeared to be somewhat in awe of what he estimated to be a twelve-hundred-pound bear, but like Lewis and Clark nearly a hundred years before him, Roosevelt thought nothing of shooting it as he looked into the bear's "evil eyes." After killing the grizzly, he triumphantly declared, "I felt not a little proud, as I stood over the great brindled bulk." Roosevelt made conservation part of the American lexicon, but even he viewed grizzlies as "evil" and their populations continued to decline even after the idea of "wise use" had solidified itself in the American lexicon.

Grizzlies had occupied the entire western half of the continent from Alaska and Canada south through Mexico, but less than a century after Euro-Americans arrived, these populations began going extinct. In 1890 the last grizzly bear was killed in Texas, followed by California in 1922, and Utah in 1923. In 1922 the U.S. Biological Survey, announced its intention to protect all bears that were not stock killers, and in 1928, Arizona limited grizzly bear hunting to a specifically demarcated season. Still, the wave of state-by-state extinctions continued and raised little concern among the American public. Hunters killed the last bears in Oregon and New Mexico in 1931; and despite its regulations, Arizona killed its last bear in 1935. By the 1960s, poison targeting wolves helped kill the last grizzlies in Mexico, and a few years after the Endangered Species Act passed, Colorado's last grizzly died from a hunter's bullet in 1979. Estimates for the continent's grizzly bear population had ranged from fifty thousand to one hundred thousand, but by the middle of the twentieth century less than 1 percent of its former population occupied less than 2 percent of its former range.[13]

From Lewis and Clark's expedition through to the early decades of the twentieth century, Americans' relationship with grizzly bears reflected their relationship with the natural world in general. Before Theodore Roosevelt and Gifford Pinchot introduced the tenets of conservation to the country, few Americans possessed what modern observers would identify as a land ethic. As westerners' attitudes toward grizzly bears demonstrated, Americans sought to tame the land without consideration of their ecological impacts. Men like Henry David Thoreau, George Perkins Marsh, and John Muir advocated maintaining a balance between human progress and the natural world, but their messages largely went unheard and did little to influence government policy. Pinchot's "wise use" conservation message, on the other hand, quickly found its way into mainstream policy.

In opposition to Muir's preservationist message, conservation did not protect natural resources such as forests, rivers, and wildlife merely so that people could feel a connection with them. Rather, they were conserved in order to be used to benefit humankind. The philosophy dictated efficiency, which meant harvesting trees and game animals in ways that were sustainable so that future generations could do so as well. No one person or group had the right to rob the public of a particular resource, because it belonged to everyone. However, because Americans viewed grizzly bears and other predators as a threat to progress and economic stability, conservationists did not attempt to protect them and in fact encouraged their destruction. In this way, conservation was a utilitarian, economic model of land management that advocated protecting the land only insofar as was necessary to reap its rewards. The creation of national parks represented a preservationist ideal championed by Muir that would become the dominant ethic by the 1960s and 1970s, but at the beginning of the twentieth century, conservation's probusiness philosophy made it far more appealing to most Americans.[14]

In the Bitterroots, the story was no different. While grizzlies were once plentiful, they had been all but eradicated from the region by the mid-twentieth century. Settlers primarily valued the region for its extractable resources, and grizzly bears inhibited their ability to pursue those industries and therefore had no place in the region's future. The Bitterroot Valley had been one of the earliest regions of Montana to be settled when Jesuit missionaries arrived in the 1840s, and even earlier, trappers had documented interactions with grizzlies in the region. In 1836 two pioneers reported encountering a grizzly while hunting along the Bitterroot River. Around the same time, a grizzly attacked a trapper and mountain man named Lawrence Rence, also known as Lolo, while he was hunting along Grave Creek in the Bitterroots. In 1852 another trapper, who had lived among the Nez Perce, encountered a grizzly and became the first known Euro-American to be killed by the great beast in the Bitterroot region.[15]

While grizzlies may have sometimes had the upper hand in these early interactions, the Bitterroots quickly became a favored spot for grizzly bear hunters. This was the case, in part, because Bitterroot grizzlies relied heavily on seasonal salmon runs from the Pacific. Congregated around favored fishing holes, grizzlies became much easier targets for hunters than if they had been dispersed, foraging for food. In the 1870s a trapper named Johnny Richie reported killing nine grizzly bears in only two weeks in the upper Bitterroot Valley. In 1893 the Carlin hunting party wounded a sow with cubs outside of Elk City, Idaho, in the heart of the Bitterroots. Another hunter, Wes Fales, who lived in the region around the turn of the century, claimed to have trapped two bears in a day on three different occasions in the region. Around this time, Fales also estimated that five to nine other trappers were active in the Bitterroots. From these numbers, Bud Moore, a Forest Service ranger on the Clearwater National Forest, later deduced that around the turn of the century, trappers killed twenty-five to forty grizzlies each year in the Bitterroot Mountains.[16]

Even though grizzly hunting was already prevalent, the historic wild fires of 1910 that swept across Montana and Idaho accelerated the bear's destruction in the region. By burning the thick, old-growth forests, these fires transformed the landscape and made the Bitterroots well-suited for grazing livestock. Ranchers began taking their stock into the mountains en masse, and by 1935, more than thirty-five thousand sheep grazed the Bitterroots. Although ranchers, along with other settlers, had previously targeted the bears and did not need any excuse to shoot one, the increased interactions between bears, humans, and livestock created additional incentive and opportunity to kill them. In 1919 the federal government established the Selway Game Preserve, but because Americans still considered grizzlies to be varmints, undeserving of the safeguards afforded by conservation's ideology, the bears did not benefit from its protections. And while grizzlies browse a range of food sources, the Bitterroot population had evolved to become so dependent on salmon that when dams, built in the 1920s near Lewiston, Idaho, began obstructing seasonal runs, the remaining population suffered greatly. As a result, they started moving down into the populated valleys to find food, which led to more contact with humans and further hastened their elimination. By 1926 the last grizzly bear was killed in the Clearwater National Forest, and the rest of the range was not far behind in losing its keystone predator.[17]

Most people living in the Bitterroot Valley likely celebrated the eradication of the grizzly from the region, but for at least one man who spent his life in the Bitterroots, the loss of the bear was tragic. Bud Moore was born in the town of Florence, Montana, in 1917, not far from Lewis and Clark's Traveler's Rest campsite. He was raised hiking and hunting throughout the northern Bitterroot region, and as soon as he came of age, he took a position as a ranger on the Clearwater National Forest's Powell District, where he worked for many years. Although he was an avid hunter who killed a grizzly bear in his youth, like others before him, Moore eventually became a dedicated environmentalist who wanted to preserve all the species of the Bitterroots, especially the grizzly. In his 1984 book *The Lochsa Story*, Moore recounts his early days as a child and young man in the region and argues for a land ethic that values biodiversity. Moore discusses the grizzly bear at length and laments the symbolic and real loss of the region's grizzly population.

As a child Moore remembered seeing at least five grizzly bears at close range, but by the time he became a ranger in the 1930s, the bears were practically extinct from the region. In 1932 Moore found nine bear scalps nailed to a tree in a meadow in his ranger district, which sadly proved to be the last confirmed visual sighting of a grizzly bear in the Bitterroots. He left the region to fight in World War II but immediately returned home and went to work for the Forest Service again after the war. In 1946 he came upon the track of a grizzly, which would be the last confirmed sign anyone would see of the great bear in the Bitterroots for more than sixty years. Every year while working as a ranger at Powell, Moore kept annual estimates for the populations of each species in his district. For a number of years after seeing that track, Moore stubbornly estimated the grizzly population at five, because he refused to admit they were gone. After a number of years, however, he forced himself to accept the regrettable truth. In an interview some years later, Moore remembered that "something died in me when I made my annual wildlife report and added a zero next to the grizzly bear." In that same interview, Moore mused over the possibility of grizzlies returning to the Bitterroots, a goal for which he spent much of his life fighting, and declared, "Oh, what grand day that will be." Unfortunately for Moore, he died at age ninety-two without seeing this dream fulfilled.[18]

Moore's story aptly demonstrates the larger narrative of the country's transitioning environmental ethic. As a young man, he had subscribed to conservation and even killed a grizzly himself, but in his later years, his relationship with the natural world had evolved, and he developed a more sensitive connection with every element of his beloved Bitterroots. Similarly, conservation remained part of mainstream American politics throughout the first half of the twentieth century, but in the post–World War II era, environmentalism quickly surpassed conservation as the dominant ethic. For the first time in nearly a decade and half, Americans had disposable income, and once gas rations were lifted after the war, they hit the road to tour many of the country's national parks and forests. Equipped with higher-quality automobiles, Americans started taking vacations that had not been previously possible. As an increasing percentage of the population came into contact with these special places, the desire to ensure their continued existence abounded even if the increased use these places received slowly eroded their splendor. The advent and subsequent proliferation of color photography, nature films, and wildlife shows also fostered a love of and desire to protect places and animals with which Americans had not previously had substantial contact. Increasingly, the country's justifications for preserving these resources relied less on economics and more on the aesthetic and intangible benefits they provided.[19]

Aiding the evolution of the country's environmental ethic, people began to move out of cities and into suburbs. This migration led them to place a higher value on quality of life and natural beauty. Coupled with an emerging passion for outdoor recreation, Americans started finding ways to incorporate the natural environments they encountered on their annual retreats into their everyday lives. On top of this, technology made enormous strides in the postwar years, and within two decades, people began to realize the harmful effects that chemicals, pesticides, and certain manufacturing by-products had on the environment. Furthermore, scientists started developing ecological theories that dictated caring for the country's diminishing natural resources on a larger, ecosystem scale. As a result, Americans not only expanded their conception of what it meant to protect the natural world and its wildlife beyond wise use, but they had the ecological theories and scientific know-how necessary to contextualize these ethics within a broader context and thus afford them greater meaning.[20]

As Americans began to make physical connections with the natural world on a larger scale, they in turn developed a land ethic that went beyond Pinchot's wise-use conservation. In his posthumous best seller *A Sand County Almanac*, renowned ecologist Aldo Leopold laid out what would become the ideological foundation of the environmental movement. According to Leopold, conservation focused "solely on economic self-interest and is hopelessly lopsided. It tends to ignore, and thus eventually to eliminate, many elements in the land community that lack commercial value, but that are (as far as we know) essential to its healthy functioning," and he encouraged people to examine the environment through a moral lens. In one of the book's most famous passages, Leopold states, "A thing is right when it tends to preserve the integrity, stability, and beauty of the biotic community. It is wrong when it tends otherwise." More than any other single statement, this would become the philosophy that would drive environmentalism over the subsequent decades.[21]

Similarly, Leopold believed that humans should be merely citizens of the land, not its conquerors. Although he eventually championed this ethic late in life, Leopold's first job was for the Forest Service in New Mexico, where one of his chief duties was killing wolves. In another of *A Sand County Almanac*'s best-known passages, Leopold recounted the death of one particular wolf and lamented, "I thought that because fewer wolves meant more deer that no wolves would mean hunters' paradise. But after seeing the green fire die [from the wolf's eyes], I sensed that neither the wolf nor the mountain agreed with such a view." With this statement, Leopold not only inspired thousands, but remarkably, challenged Americans to include even the most hated predators under environmentalism's growing umbrella.[22] While grizzly bears had been subjected to the same predator control programs that encouraged Leopold to shoot the green-eyed wolf, by the time Americans were reading *A Sand County Almanac*, they had started to value the presence of grizzlies on the landscape; and passages like this one reaffirmed this new appreciation.

Along with breakthroughs in biology and ecology, romantic conceptions of nature's invigorating effect on the human spirit also helped shape this ethic. Writers and thinkers such as Wallace Stegner and Edward Abbey revived the message of earlier nature lovers such as Henry David Thoreau and John Muir and contended that wilderness shaped the American character and was necessary to maintain it. These environmental philosophers came to believe that the environment, in its entirety, maintained the human soul and was necessary for our continued existence. In one of his most famous lines, Stegner waxed, "Something will have gone out of us as a people if we ever let the remaining wilderness be destroyed." Even Secretary of the Interior Stewart Udall, who served under Presidents Kennedy and Johnson noted, "The conservation movement . . . was itself disorganized and outdated." The advances made in biology and ecology provided a basis for the more romantic and emotional appeals being made for an expanded land ethic, and combined with legislative action, these forces would define the environmental movement over the rest of the century.[23]

In this same vein, many people started to view predators, such as wolves and bears, as the ultimate symbols of wilderness, so while conservation mandated their destruction, environmentalism necessitated their preservation. By the 1940s, grizzly bears occupied less than 2 percent of their former range in the continental United States and their population had dropped even lower. Many western land users, such as ranchers, continued to seek the bear's destruction, but at a national level, Americans started to recast the grizzly as a national treasure.

In the postwar years, congruent with America's increasing appreciation for the nation's natural wonders, the bear enjoyed a surge in popularity resulting not only from the growing tide of the environmental movement, but more specifically from increased visitation to the western national parks. As people became acquainted with bears in a new context, admiration and respect replaced fear and hatred as the emotions most readily associated with grizzlies. Grizzly bears had always been popular among visitors in western national parks, but as more people flocked to these pleasure grounds in the years following World War II, grizzly bears' public image received a substantial makeover that reverberated throughout the country.

Yellowstone National Park had been one of the country's crown jewels ever since it became the first national park in 1872; and by the 1930s, it was one of the last best strongholds for grizzly bears in the country. During this era, the National Park Service maintained bear feeding grounds in the park where tourists gathered daily to watch grizzlies feed on hotel garbage. Each of the feeding grounds featured a central stage, surrounded by bleachers and every night, hotel guests would file into the seating area following their evening meal. After everyone was settled, hotel staff would bring out the kitchen's garbage, spread it out on the stage below, and watch as grizzlies, who were accustomed to the nightly routine, came out of the nearby woods to feast.[24]

The shows were enormously popular as they afforded tourists an easy opportunity to view the bears from close range, but by the end of the 1930s, park authorities wished to close these feeding grounds because they detracted from the natural experience the park tried to create. Because of their popularity, however, the Park Service was hesitant to deprive visitors of this experience, which was often a highpoint. World War II provided the Park Service the necessary distraction to close the feeding grounds, but when tourists returned, they found new ways to interact with the park's bears at close range. Although feeding bears from the roadside was always against park rules, tourists liked doing it so much that the Park Service enforced the rule only sporadically. Feeding bears directly from car windows became so common in fact that bears congregated along roadsides in droves, and these interactions replaced the nightly feeding shows as one of the main attractions for tourists coming to Yellowstone. Families would commonly stop before entering the park to pick up Twinkies or other similar foods, with the explicit intent of feeding them to roadside bears. Minor injuries occurred frequently, but these incidents made little impression on tourists who were anxious to take part in an essential Yellowstone activity.[25]

These new interactions became especially significant because park attendance after the war proliferated, bringing many more Americans into contact with the great bear. Before the war, park attendance remained relatively low and was accessible only to the wealthy who could afford the expensive train ride and typical five-day tour through the park. In the immediate postwar years, however, visitation exploded. In 1939 only 486,000 people visited Yellowstone. By 1949 that number had grown to 1.1 million, and twenty years later, in 1969, annual visitation had reached nearly 2.2 million. [26] American families were hitting the road like never before, and when they came to Yellowstone, seeing a grizzly was almost always the highlight of their trip.

As people became familiar with bears in this seemingly tame context, grizzlies went from hated predator to beloved national treasure, and new images of bears in popular culture accelerated this transition. Just as the war was ending in 1945, before Yellowstone's roads became choked by mile-long "bear jams," the Forest Service unwittingly took the most significant step in remaking the popular image of bears. That year, the agency introduced its new spokesman, Smokey Bear. Although Smokey was a black bear, his instant appeal had lasting positive effects for grizzlies as well. In a few short years, Smokey's fire safety message made him one of America's most recognizable and beloved characters. A few years after the Forest Service introduced Smokey to the public, a group of fire fighters rescued a black bear cub that had been orphaned by a man-made forest fire in New Mexico. The cub was taken to the National Zoo in Washington DC where he acquired the name Smokey. Not only did the Forest Service's Smokey now have an adorable, real-life identity, but the "real" Smokey came attached to a story that inspired Americans to protect bears as well as prevent forest fires. Within a few short years, Smokey's face would become fodder for comic books, stuffed animals, television shows, children's toys, and many other emblems that encouraged Americans to think of bears as friends in need of protection, not fearsome predators whose existence challenged the American way of life.[27]

A few years later, Yogi Bear, the picnic-basket-stealing cartoon bear, hit television screens across the country, joining Smokey in the nation's unwitting campaign to recast grizzly bears as affable companions, not blood-thirsty killers. Yogi, like Smokey, became a national sensation, and the Park Service even incorporated him into a few of their bear safety campaigns. Although the message that bears were harmless was far from prudent and undoubtedly increased incidents between park visitors and bears, the influence that Yogi and Smokey had on how Americans perceived the great bear was ultimately beneficial.

When Disney released *Bear Country* in 1953 as a part of its True-Life Adventure series, bears' public image profited once again. The hour-long documentary followed a single family of bears over the course of a year in a way that anthropomorphized them and cast them as sympathetic creatures in need of human protection. By the late 1960s, media attention for bears transitioned to live action drama with human actors interacting with real bears. From 1967 to 1969, CBS aired *Gentle Ben*, a half-hour-long television show in which a tame black bear played helpmate and friend to a game warden and his family. The feature-length film *The Life and Times of Grizzly Adams* (1974), and the spin-off television series (1977–78) by same name, starring Dan Haggerty, also reinforced the message that bears were amiable creatures and companions. Not only were bears not threatening, but they were potential friends and helpmates that deserved protection. Not surprisingly, all this positive media enabled grizzlies to transition quickly from irredeemable killers to beloved national icons.[28]

Along with receiving an improved status in the popular imagination, in the postwar years the scientific community dedicated new interest to the grizzly as well. Biologists, who had previously devoted little serious attention to grizzlies, developed a new appreciation for the great bear and research on the bear flourished. Two of the most influential bear biologists in this new wave were twin brothers, John and Frank Craighead, who began researching Yellowstone's grizzly bear population in 1959. Focused primarily on the bear's reproduction and mortality rates, the brothers' findings proved groundbreaking. Consequently, as the 1960s progressed, scientists and government officials became better equipped to manage and protect the bears.[29] However, it was an event in Glacier National Park that became the catalyst for a new era of bear management.

On a single night in 1967, known as the "Night of the Grizzly," two different bears killed two people in entirely different regions of the park. Believing these attacks had resulted from lenient bear management policies that gave bears ready access to human food and encouraged them to associate people with an easy meal, the Park Service in Glacier and Yellowstone decided to crack down on these perpetrators in an effort to "re-wild" the bears. This meant closing every garbage dump and ending lax enforcement of roadside feeding bans. The Craigheads recommended that Yellowstone phase this policy in slowly for the sake of the bears, which had become reliant on human food, but the Park Service wanted it completed as soon as possible. Despite vocal protestations from the Craigheads and their supporters, the Park Service forced the bears to go cold turkey. As a result, bears that had grown accustomed to receiving handouts found themselves out of luck and in trouble when they entered campgrounds looking for picnic baskets. Between 1970 and 1971, eighty-eight bears in the park and the surrounding ecosystem died or were removed by the Park Service, and over the next few years, the grizzly population plummeted. By 1975 biologists estimated that Yellowstone's grizzly population had dropped to as few as forty bears.[30] Luckily for the bears, public opinion and new legislation had finally combined to make their protection a reality.

By 1975 the environmental movement had reached its peak after more than a decade of legislation inspired by its ideals had brought stricter regulations governing Americans' relationship with the natural world. But before environmentalists had the power to inspire, craft, and influence legislation, Rachel Carson's 1962 book *Silent Spring* inspired the country to action and helped bring environmentalism to mainstream America. According to historian Shannon Petersen, "If the environmental movement had a beginning," it was with Carson's book. In the book, Carson, who had trained as a biologist, examined the negative effect that the country's liberal use of pesticides had on its wildlife. Within the next few years, legislation related to the environment and wildlife proliferated as did public support for these actions. In 1964 the Fish and Wildlife Service took the first step toward expanding protections for wildlife and formed the Committee on Rare and Endangered Species, which produced a list of sixty North American species that faced extinction. The following year Congress debated a law to protect these species, but another two years would pass before it would adopt any such law.[31]

That moment came in 1966, under Democrat Lyndon B. Johnson's administration, in the form of the Endangered Species Preservation Act. While ineffectual, it was the federal government's first attempt at protecting endangered species on any broad scale and was a turning point in the federal government's relationship with wildlife. The act was significant because it established a precedent that the federal government, not the states, would control endangered species recovery. However, the act itself was relatively weak. First, it was only enforceable within national wildlife refuges, and it applied only to vertebrates. Secondly, it did not protect species habitats, so it did nothing to prevent actions that would harm endangered animals indirectly. Finally, it charged the Departments of the Interior, Agriculture, and Defense with enforcing the law only "insofar as is practicable." This ambiguous, nonbinding, language made the law difficult to uphold because all participation was essentially voluntary. Still, the 1966 act signaled the nation's growing concern for protecting wildlife and provided a model for subsequent legislation.[32]

Almost immediately after the passage of the 1966 act, Congress began considering amendments or new laws to strengthen endangered species protection. Public support continued to swell, and three years after the first act, Congress passed the Endangered Species Conservation Act of 1969. This act extended protection to amphibians and reptiles, charged the U.S. Fish and Wildlife Service with compiling a global endangered species list, and prohibited the importation of any endangered species.[33] Although the 1969 act made improvements to the 1966 law, it left some major problems untouched, and it, too, was relatively powerless. However, by extending protection to species other than charismatic megafauna, the federal government embraced a more expansive form of biocentrism and recognized the value of species not widely appreciated.

In addition to endangered species laws, the 1960s saw a proliferation of environmental legislation including the Wilderness Act, the Clean Air and Clean Water Acts, the Wild and Scenic Rivers, and the National Environmental Policy Act, but the public's appetite for environmental legislation had not yet been satiated. In 1970 Republican president Richard Nixon, speaking to the Council on Environmental Quality, announced that the country could no longer afford to treat land as an infinite resource, and he encouraged Americans to increase their knowledge of environmental issues as legislation would not be the only solution. That same year, the country celebrated its first Earth Day, and in 1972, Congress passed the Marine Mammal Protection Act (MMPA), which extended protection to a greater range of species. Connected with the MMPA, the United States hosted the Convention on International Trade in Endangered Species the following year, which brought more than eighty countries to the United States to discuss the extinction crisis.[34] By all measures, at this moment in history the environmental movement was an unstoppable political force.

Faithful adherents of conservation, who believed in it strongly as an economic tool, did not know how to react to the environmental movement's more far-reaching philosophies and goals. Although they initially supported them, they increasingly found their views at odds with those of the movement's swelling mass. Loggers, ranchers, farmers, and miners made a living from extracting the resources that environmentalists wanted to protect, and although they were not necessarily against the reforms being proposed, these actions impacted their economic livelihoods much more so than for the average American living in an urban or suburban setting. Even so, the environmental movement was so popular that the political environment prevented them from making a stand against it, and they were forced to bide their time.[35]

In the meantime, the environmental movement's popularity continued to soar, and in a 1972 address, President Nixon declared that even the most recent legislation "simply does not provide the kind of management tools needed to act early enough to save vanishing species." As a result, he called for "a stronger law to protect endangered species." A year later, after little debate, the Senate passed the Endangered Species Act of 1973 by a vote of ninety-two to zero, and the House followed with a vote of 390 to 12. On December 28, 1973,

President Nixon signed the bill into law.[36] As backing for the law was nearly unanimous, Congress's actions confirmed that the environmental movement had a mandate that few national political leaders were willing to challenge.

In Montana and Idaho, where the question of grizzly recovery would rage two decades later, support for the ESA was representative of the nation at large. Montana's entire delegation, Sen. Mike Mansfield (D), Sen. Lee Metcalf (D), Rep. Dick Shoup (R), and Rep. John Melcher (D), voted in favor of the act. In Idaho, support was not unanimous, but still solid. Both of its senators, Frank Church (D) and Jim McClure (R), voted in favor, while Rep. Steve Symms (R) cast one of the twelve votes against. The final member of Idaho's delegation, Orval Hansen (R), did not vote.[37] Considering that both states depended on extractive industries, the fact that six of seven voting representatives voted in favor of the measure indicates how widespread support for environmentalism and endangered species legislation was in the early 1970s. Environmentalism's power may have been fleeting, but for the moment, it was substantial.

The Endangered Species Act of 1973 went far beyond the previous endangered species laws by strengthening not only its protection of species but their habitats as well. What earlier iterations of the law missed was that harming a species habitat was a less direct way of harming the species itself. For the first time, the 1973 act recognized that protecting habitat was essential to protecting wildlife. Additionally, the new law gave the Fish and Wildlife Service responsibility for listing and delisting species and created two possible classifications for listing: endangered, for species faced with imminent extinction; and threatened, for species that could become endangered in the foreseeable future. Also, section 4 of the law stated that science would dictate the decisions for listing and delisting, and it made cooperation at all levels—federal, state, and private—mandatory. Most importantly, the act stipulated that no listed animal could be killed, trapped, harassed, and so on. In what became one of the most controversial sections of the act, section 7 essentially barred any federal agency from funding or taking any actions that would threaten listed species or the habitat deemed "critical" to their existence.[38]

According to environmental historian Roderick Nash, the passage of the ESA was a landmark event that signified the evolution of a belief that animals possess certain natural rights. In his book *The Rights of Nature*, Nash compares the ESA's implications for animals with the Declaration of Independence's significance to the American colonists, the Emancipation Proclamation's impact on African Americans, and the Nineteenth Amendment's bearing on women, to name a few. The book assumes a highly laudatory tone as Nash considers the passage of the act the culmination of a centuries' long process in which animals slowly gained more rights. Additionally, Nash treats the passage of the ESA as proof that Americans, across the country, widely recognized that animals possess inherent natural rights.[39]

Much of Nash's argument is reflected in a series of surveys conducted by Yale University researcher Stephen Kellert, beginning in 1979, that gauged Americans' attitudes toward animals—domestic, wild, and endangered—and how those sentiments measured up against other values. Not surprisingly, 89 percent of respondents to one survey favored protecting bald eagles, even if it meant increasing costs to an energy development project; but perhaps less expected was the 73 percent of respondents who favored protecting mountain lions under the same set of factors. Additionally, the survey concluded that 56 percent of people would support a project that set aside land for grizzly bears, even if it hurt the timber industry; and among respondents from the Rocky Mountain West, that number increased to 59 percent. Similarly, 76 percent of respondents agreed that tree harvesting should happen in a way that does not harm wildlife, even if it costs more; and only 36 percent of the public agreed that environmental goals were a threat to the country's economic prosperity.[40]

Although many of the responses seemed to indicate the overwhelming popularity of the environmental movement and a wholesale shift in the nation's regard for animals, the transformation was not as complete as it initially appeared. Forty-four percent of people agreed that natural resources must be developed even if wilderness and wildlife populations decreased as a result; and 76 percent of sheep ranchers and 82 percent of cattle ranchers believed that environmental goals *were* a threat to the country's economic prosperity.[41] Both of these numbers were signs that while the environmental movement had certainly reshaped Americans' opinions of animals, Nash's estimation of the act's implications was greatly inflated as the older conservationist ethos lingered in the American mind.

Although the passage of the act was a landmark moment, enforcing and fulfilling its promises were not automatically assured. As Shannon Petersen argues, in *Acting for Endangered Species*, because Congress did not debate the act heavily and because it did not receive much press, it was not the grand symbol that Nash believed it to be. Petersen is not alone, as other scholars have also noted that the act was considered merely a symbolic gesture. Neither Congress nor the public immediately understood its power or significance, and many probably did not believe that the act meant recognition of natural rights for animals.[42]

Furthermore, the act would prove to be a highpoint in the environmental movement rather than another step in an uninterrupted march toward achieving environmental utopia. As environmental historian Hal Rothman explains, the environmental movement was the product of a society overflowing with "affluence, abundance, prosperity, and optimism," and did not represent a permanent ideological transformation. As the Watergate scandal, the conflict in Vietnam, and a stagnant economy all came to define the 1970s, the environment took a backseat to what many Americans believed to be more pressing issues. More fundamentally, historian Paul Sutter argues that environmental ethics are reworked on a generational basis to fit the needs of a particular time period. As opposed to evolving in a constant and linear fashion, Sutter's model posits that they evolve sporadically and are vulnerable to moving backward as well as forward.[43] The Endangered Species Act and the environmental movement were the products of a distinct generation and a distinct set of problems; but by the mid-1970s, the forces of history were ushering in a new era that was less receptive to regulation and far less will-ing to make sacrifices for the environment or endangered species.

Nevertheless, the passage of the act was undoubtedly a landmark *legislative* victory for both environmentalists and wildlife, even if it was not the product of a newly evolved, universally held ethic. The environmental movement's power would wane over the succeed-ing years, but the ESA afforded the federal government and environ-mentalists the necessary legal tools to fight species extinction. For predators, the ESA was especially significant. Not only did it fulfill Leopold's vision and acknowledge their inherent right to exist, but it was the culmination of a wholesale reversal in federal policy—from eradication to protection.[44]

Similar to William Wright's relationship with the grizzly bear, the nation's attitudes toward the environment changed dramatically over the twentieth century. At the turn of the century, the conservation movement took the reins of natural resource policy, protecting the land and its wildlife as far as was practicable for human benefit. Their attempts were not motivated by a desire to preserve the land for its own sake, but for the sake of subsequent human use. By the 1960s, Americans faced new environmental challenges that conservation was not equipped to handle, and federal natural resource policies adapted to combat these new threats. This shift ushered in an ethic that justified protecting wildlife not merely in terms of cost-benefit analyses, but for the ecological roles animals filled and the positive experiences they provided humans.

The switch from conservation to environmentalism helped reshape Americans' opinion of the great bear in substantial ways that not only made the bear a popular figure in the public mind, but also led to significant shifts in policy. As Americans had new opportunities to see bears up close in national parks as well as on their television screens, they began to view bears as friendly, jovial, and harmless rather than as the menacing, man-eating beasts that earlier generations of Americans considered them. Aided as well by advances made in ecology and biology, the grizzly became a national icon representative of our ties with open land and wild spaces, so that many Americans viewed protecting the bear as an essential part of preserving their own heritage.

Unfortunately for the grizzly, the environmental movement's nearly unanimous support had evaporated within the next decade as supporters and detractors began to divide along partisan lines, and many Americans rejected the implications of the Endangered Species Act. This breakdown of consensus, and the attempt by some groups to resurrect Pinchot-era conservation policies, would prove that environmentalism had failed to reshape Americans' priorities at a foundational level, bringing many of the movement's achievements under scrutiny and undermining any expectation that the reintroduction of grizzly bears could win unanimous support.

2

Endangered Species, Environmental Politics, and the American West

By the 1990s Catron County, New Mexico, was in trouble. Its economy, which relied heavily on extractive industries, was in a steep decline, and the resources on which those industries depended fared even worse. Decades of overuse had left the high, arid forests of the Gila National Forest a ghost of what they had once been. Grasslands had been overgrazed, forests overlogged, and as a result, spring snowmelt quickly drained from the hillsides. With no vegetation to trap the water, the rivers and streams surged to previously unseen levels, erosion increased, and the land was in the process of being wholly transformed. Fish and wildlife populations suffered as a result, and the region's defining character was being undermined after a hundred years of taking from the land without consideration for the long-term effects. When the U.S. Forest Service attempted to rehabilitate and restore the land by limiting logging and supporting wildlife populations, the agency quickly became the target of local land users who declared the county under siege by the federal government.

By co-opting a federal land law intended to protect Native American heritage sites, Catron County claimed the federal government was not respecting its "culture and customs" as the law required. Rewriting the county's own laws, Catron essentially invoked a nullification argument in which the county had the right to trump federal laws with which it disagreed. At the symbolic center of this debate was Kit Laney, a local rancher who owned just a few hundred acres, but grazed his cattle on the hundreds of thousands of acres of Gila National Forest land that bordered his property. When Laney stopped paying grazing fees, he became a folk hero and champion of this second iteration of the Sagebrush Rebellion known as the Wise Use Movement (also the County or Property Rights Movements). Laney exemplified the romantic cowboy ideal and embodied the rugged individualism and self-sufficiency that Wise Users and Old Westerners alike championed.[1]

Dozens of other counties across the West followed suit, and some leaders of the movement advocated open violence. On July 4, 1993, Dick Carver, a county commissioner in Nye County, Nevada, drove a bulldozer at a U.S. Forest Service ranger in hopes of reopening a dirt road the federal government had recently closed. With a copy of the U.S. Constitution in his chest pocket, Carver pointed the bulldozer at the ranger as a way of asserting his firmly held belief that the federal government had no right to administer land within Nevada or any other state. Carver's joy ride was not an isolated incident. A few years later, a Forest Service office in Carson City, Nevada, was bombed, and in many places across the rural West, Forest Service rangers feared for their lives.[2] While the majority of these incidents unfolded more than a thousand miles away from the closest wild grizzly bear, the backlash against federal environmental regulations was widespread across the West; and Old Westerners in Montana and Idaho shared an ideological connection with and drew inspiration from those leading the Wise Use Movement in Nevada and New Mexico.

By the 1970s, the environmental movement had reached full swing. It had taken control of federal natural resource policy; and the passage of the Endangered Species Act, along with the listing of the grizzly bear as threatened, in 1975, exemplified the movement's success. However, it proved to be a high-water mark. Throughout the 1960s, the environmental movement had been nonpartisan, but by the end of the 1970s, a conservative tide had swept the nation that left the Republican Party generally opposed to environmentalism's reforms, while Democrats continued to support them. In the West, this backlash had its own unique twist. The region had long been a bastion of progressive and experimental politics, but as extractive industries faced threats from outside influences, the environmental movement became an easy target that extractive land users saddled with much of the blame for their difficulties. As a result, the West, as a whole, became more conservative, with the Sagebrush Rebellion epitomizing this shift. This so-called rebellion, which was made up of Old West ideologues who fought to revert the country back to the wise-use model of conservation, resisted many of the environmental movement's reforms, and the Endangered Species Act was one of its favorite targets. The rebellion demonstrated that environmentalism, especially as it pertained to wildlife, was not permanently ingrained in the American psyche and that its legacy would have to rely on its legislative accomplishments. The rebellion fizzled by the mid-1980s, but many of its sentiments persisted, so that by the time the Fish and Wildlife Service and bear advocates were taking meaningful steps toward recovering grizzlies in the Bitterroots in the early 1990s, the country's environmental conscience was starkly divided.

Two years after Congress passed the Endangered Species Act, the Fish and Wildlife Service formed the Interagency Grizzly Bear Study Team to determine the condition of the grizzly bear population across the continental United States. After compiling and analyzing the research of a number of bear biologists, the team recommended the grizzly bear for listing under the Endangered Species Act of 1973 as a threatened species, and on July 28, 1975, the Fish and Wildlife Service added the bear to its growing list.[3]

As the ESA stipulated, the Interagency Grizzly Bear Study Team based its decision to list the grizzly bear on a five-factor analysis, and they determined that grizzly bears faced threats from each factor except disease and predation. According to the team, destruction of habitat, overutilization for commercial and recreational purposes, lack of existing regulations, and other man-made factors all compromised the bear's future existence. This meant that the bear could not be taken off the list (delisted) until all four threats were resolved. Of those four factors, loss of habitat was the most pressing. Grizzly bears require huge amounts of relatively intact habitat to survive because they cover immense areas annually and throughout their lifetimes. The more they come into contact with humans, whether on roads or private property, the less likely they are to survive. Typically, male grizzlies cover more land than females, but females' ranges are not insignificant. On average, a female usually covers about 150 square miles annually and up to 340 square miles in her lifetime, while a male grizzly bear will roam across more than 300 square miles each year and about 1,450 square miles in his lifetime. By 1975 grizzly bears' ability to roam in the lower forty-eight states was limited as they persisted only in island populations in Montana, Wyoming, Idaho, and Washington. And even in these last strongholds, logging, mining, new road construction, expanding towns, and the presence of livestock increased habitat fragmentation, decreased its quality, and seriously threatened bears' long-term viability.[4]

The USFWS also decided that bears faced threats from certain recreational activities, lack of existing regulations, and "other factors" influencing their future existence. In the 1960s, Montana and Wyoming continued to hold seasonal grizzly hunts, and outside of hunting season, ranchers continued to shoot any bears they found near their stock. According to the Craigheads' research, 47 percent of grizzly bear mortalities resulted from human hunters. Within "other factors," the Fish and Wildlife Service cited the lack of knowledge concerning carrying capacities for certain ecosystems, reproductive behavior, and overall population trends as reasons for listing the bear. Since then, biologists have discovered that grizzly bears reproduce slowly. At the earliest, females begin reproducing at age three, but this can often be delayed until they are as old as eight. Additionally, females reproduce only every three years, as they care for each litter for two years and do not mate while caring for their cubs. Healthy females may have as many as four cubs in a litter, but two is most common, which means they cannot sustain high mortality rates. Finally, the USFWS determined that motorized recreation, ranching, mining, and logging close to grizzly populations was harmful because it increased bears' contact with humans, which often led to their removal or death.[5] For grizzlies to be taken off the endangered species list, managers would have to mitigate all of these factors.

As well as stipulating the reasons for a species' threatened status, the Endangered Species Act requires that species' distinct populations be listed and delisted separately. Adhering to this provision, the USFWS listed grizzlies as threatened in the four distinct island populations in which grizzlies resided, so that each population could be managed independently based on its distinctive characteristics. The four ecosystems included the Greater Yellowstone, the Northern Continental Divide, the Cabinet-Yaak, and the Selkirk, all of which were at least partially located in Montana. A few years later, the USFWS listed the North Cascade ecosystem of Washington, the San Juan ecosystem of Colorado, and the Bitterroot ecosystem of Montana and Idaho as evaluation areas that deserved further study. Biologists were unsure if grizzlies still resided in these areas, so they listed them as evaluation areas, intending to conduct further research to determine the status of their grizzly populations and their suitability to host a population of grizzlies.[6]

Because the federal government listed the grizzly bear as threatened, the bear earned important protections under the Endangered Species Act. Montana and Wyoming's grizzly hunting seasons ended, and ranchers could no longer shoot bears that attacked their livestock. Only in self-defense could someone shoot a bear. Additionally, in an effort to protect the bear's habitat, land managers were forced to treat logging, mining, and ranching as secondary interests in ecosystems in which the bear was listed.

Although the Endangered Species Act was enormously popular when it became law in 1973, and grizzly bears were able to benefit from its protections, the law's popularity was short lived, and

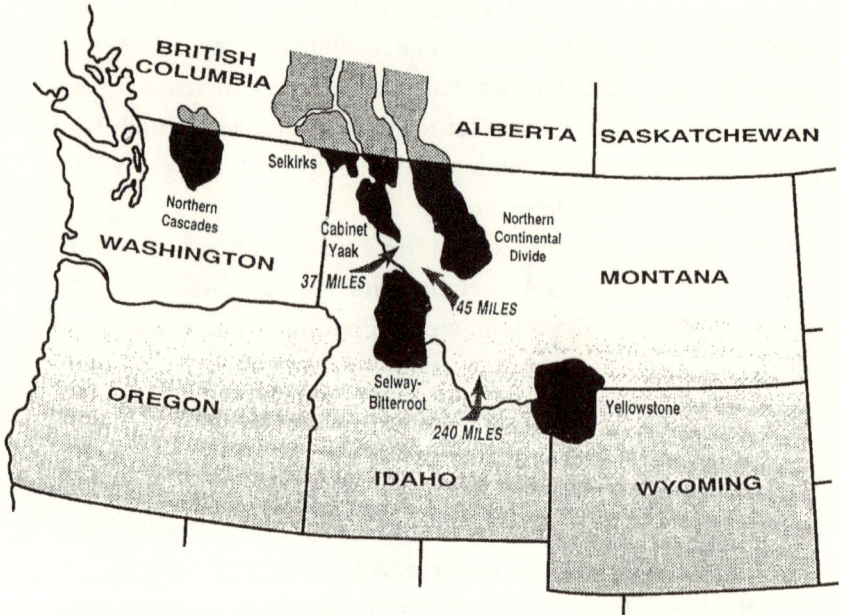

Map 2. Map of U.S. Fish and Wildlife Service grizzly bear recovery areas in the United States. Of the six grizzly bear recovery areas recognized by the U.S. Fish and Wildlife Service, only the Northern Continental Divide and Yellowstone ecosystems had healthy populations of grizzlies in the 1990s. A population in the Bitterroot ecosystem would potentially help connect these two island ecosystems. U.S. Fish and Wildlife Service.

it quickly became one of the favored targets of the environmental movement's critics. Much of this disapproval can be traced back to 1966. That year the Tennessee Valley Authority (TVA) began building the Tellico Dam on the Little Tennessee River. Environmentalists' initial attempts to stall the dam failed, but in 1973, a researcher from the University of Tennessee discovered in the river the snail darter—a small, three-inch fish that had been previously unknown to the world. Because the fish required the flow of strong currents, the completion of the dam would be its death sentence. Believing the Little Tennessee to be its only residence, environmentalists rallied, petitioning the secretary of the interior to list the darter as an endangered species. Obeying the letter of the law, the USFWS listed the snail darter as endangered in November 1975, even though the dam was nearly completed.[7]

The fish made it onto the Endangered Species List, but the head of the Fish and Wildlife Service's Endangered Species Office initially did not want to take up the cause because he believed it would threaten the law itself. Many environmentalists agreed and at first avoided the issue because the diminutive size of the fish made the case seem ridiculous. Ignoring this warning, a handful of environmental groups filed suit against the TVA in 1976 to stop the dam's completion, citing section 7 of the ESA. Section 7 stipulated that "any action authorized, funded or carried out by such agency is not likely to jeopardize the continued existence of any endangered species or threatened species or result in the destruction or adverse modification of habitat of such species." The Tellico Dam's construction undoubtedly came into conflict with this clause, but the TVA and its proponents argued that too much money had been spent on the dam by that point to abandon the project. Environmental sociologist Stephen Kellert aptly noted that the snail darter issue pitted "new and barely emergent environmental values against deeply entrenched assumptions of social worth and economic importance," and the district court ruled in accordance with the older set of values. However, the matter eventually reached the Supreme Court and in June 1978, the Supreme Court ruled six to three on *Tennessee Valley Authority v. Hill* in favor of protecting the snail darter.[8]

Forced to realize the act's power, politicians, who had no desire to see the protection of a little-known fish trump the interests of economic development, began looking for ways around the court's decision. Led by Republican senator Howard Baker of Tennessee, the Senate passed the Baker-Culver Amendment, which created the Endangered Species Committee. Nicknamed the "God Committee," by its detractors, this seven-member committee had the power to decide whether the benefits of a government project could override the courses of action needed to protect a listed species. Baker led this committee, and he wanted to ensure the dam would be completed, but the committee overruled him and found in favor of the darter. Not to be deterred, Baker and John Duncan, another Republican representative from Tennessee, attached riders to an energy bill in 1978 that allowed construction on the dam to move forward. The bill passed and the following year, the TVA closed the flood gates on Tellico Dam. The snail darter seemed doomed; but a year later, scientists found snail darters living in other rivers around the region, and a few years after that, the USFWS downgraded the darter's status from endangered to threatened.[9]

Even though the snail darter survived Tellico, the ability of conservative politicians to circumvent the law undoubtedly weakened the Endangered Species Act, and the entire episode exposed the law to scrutiny as never before. Endangered species protection previously had transcended economic interests and partisan lines, but following this episode, the sacrifices required to protect endangered species brought the principle of the idea into question. On top of that, by the late 1970s, the country's economic outlook was bleaker, and conservation's proeconomic mantra made it appealing once again. Conservationists who had not fully bought into the tenets of environmentalism no longer felt political pressure to make sacrifices for the sake of the environment, much less a three-inch fish, and putting into place environmental reforms was no longer as easy as it had been just a few years earlier.[10] More than a decade later, the attempt to reintroduce grizzlies to the Bitterroots would face resistance from a much more experienced and savvy opposition, but the snail darter episode was the first indication that the environmental movement's mandate would not last and that future efforts at endangered species protection would face concerted opposition.

As the snail darter incident was the first major endangered species controversy, opponents' arguments were relatively simple and unsophisticated. But as this issue was unfolding, the Old West was in the process of launching its own attack against the environmental movement—one that was much broader and ideologically organized. Environmentalism drove federal natural resource policy during the

1960s and early 1970s, but the industries it attempted to regulate continued to grow over this period. The population boom and mass migration to the suburbs that had begun in the 1950s and helped inspire the environmental movement also increased the need for logging, ranching, farming, and mining, much of which took place in the West. As R. McGreggor Cawley put it, "Despite growing public support for environmentalism, the nation's demand for public land resources had not decreased."[11] Extractive land users who had been some of the main proponents of conservation tried to maintain an ideological connection to the new environmental movement, but increasingly, as they found themselves the targets of environmental regulation, Old West conservationists began looking for ways to reinstate the older, less restrictive wise-use ideology.

At first this was not possible, but as the 1970s unfolded, many of the conditions that had made the environmental movement so popular began to wane. The environmental movement had largely relied on government action at the federal level, but as LBJ's Great Society faltered, the war in Vietnam grew unpopular, and stagflation set in, many Americans lost faith in big government. At the same time, the wealth and affluence that allowed Americans to devote resources to environmental issues gave way to hard times, and people lost their inclination to make economic sacrifices for the sake of the environment. As these conditions came to define the late 1970s, the environmental movement and the federal government lost much of their appeal. Realizing this, Old Westerners seized the opportunity, knowing that their criticisms had the potential to resonate widely.[12]

Tensions finally came to a head in 1979, when Nevada's Republican-dominated state legislature passed a bill announcing that the state was going to assume management over the federal lands within its boundaries because it perceived a bias in the federal government for preservation over economic development. Over the next few years, a number of western states in which the federal government controlled a substantial portion of the state's land joined the rebellion. Because the Endangered Species Act and other environmental programs were the focus of the rebellion's anger, it also attacked the federal government writ large. Sagebrush rebels denounced the federal government for trying to reform grazing allotment procedures by allowing public input, and for creating wilderness areas, which prohibited activities such as logging and grazing. Additionally, the rebellion leveled a broader accusation that the federal government did not respect state sovereignty.[13]

While the majority of the Sagebrush Rebellion's attacks were directed at the federal government, the influx of New Westerners, many of whom were environmentalists and had migrated to the region in recent years, undoubtedly inspired much of the group's anger. Unlike their older counterparts, few New Westerners relied on extractive industries; and while the reforms instituted by the environmental movement may have hurt the Old West's economy, they benefitted industries such as tourism upon which New Westerners depended.[14] So while the Sagebrush Rebellion mostly focused its attacks on the federal government, the rise of the New West was the true catalyst for this anger. If New Westerners had not flocked to the region, the West would have remained relatively removed from the influence of the environmental movement. As a result, much of the rebellion's frustration was insular, directed toward this new demographic because of how it threatened to overhaul the region's identity.

In the Northern Rockies, the Sagebrush Rebellion did not take hold to the same degree that it did in the Southwest. Wyoming was the only state in the region to pass legislation in the spirit of the rebellion, and Wyoming's legislation actually went beyond the scope of what Nevada's legislature initially passed. Nevada's bill only applied to lands administered by the Bureau of Land Management, but Wyoming's bill included Forest Service land as well. Idaho and Montana largely remained on the outside of the rebellion. Montana did pass a bill appropriating money for a multistate study related to public lands issues, but in both Idaho and Montana, true rebellion legislation was defeated. Even so, support for the rebellion thrived within pockets of both states. Idaho senator Jim McClure and Wyoming senator Malcolm Wallop endorsed the rebellion, and in all three states plenty of support existed in the more rural areas even if it did not translate into legislation.[15]

Ironically, the rebellion became a national movement when the country's top federal employee declared his support for it. In 1980 presidential candidate Ronald Reagan declared himself a Sagebrush Rebel, and after he won the election, he appointed James Watt, a prominent conservative who also sympathized with the rebellion, as secretary of the interior. Watt once remarked, half-jokingly, "If the troubles from environmentalists cannot be solved in the jury box or at the ballot box, perhaps the cartridge box should be used." With statements such as this, the Wyoming native fueled the movement's rhetoric and helped secure its presence on the national stage. For the first few years of the 1980s, the rebellion, which at times advocated open and violent confrontation with the federal government, successfully blocked environmental reforms throughout the West, but by the middle of the decade, it had withered.[16]

From the ashes of the Sagebrush Rebellion rose the Wise Use Movement, consciously adopting Pinchot-era rhetoric. In some regions, this movement became the County or Property Rights Movement, but all of these factions promoted an agenda similar to that of the Sagebrush Rebellion, although they never achieved the same degree of cohesion. In addition to adopting the principles of early twentieth-century conservation, these movements championed local control and private property rights and believed the federal government had no right to impose any sort of regulation on states. They criticized environmentalists for supporting a socialist agenda and accused them of being well-educated, upper-class snobs, out of touch with the average American. These movements were generally more moderate than the Sagebrush Rebellion, but at times, as Dick Carver and his bull dozer demonstrated, they, too, promoted violence.[17] While barely organized or formal enough to deserve the moniker of "movement," these factions retained influence throughout the 1990s and would provide the philosophical basis for the opponents of reintroduction, setting the tone for the region's political climate.

Just as the Sagebrush Rebellion was gaining momentum, another controversy brought the Endangered Species Act back into the spotlight, but this time it was in the West, and it exemplified everything the Sagebrush Rebellion and the Wise Use Movement were fighting. Logging had been essential to the economies of Oregon and Washington since the beginning the century, but by the end of the 1970s, the continuation of these activities threatened the future existence of the northern spotted owl. Only three island populations of the spotted owl existed across North America, and the northern subspecies inhabited the old-growth forests of the Pacific Northwest and depended upon them to maintain their reduced numbers. The spotted owl was not yet listed under the ESA, and the timber industry wanted to keep it that way because if it ever were to be listed—via section 7—all logging in the region would come to a virtual halt. The Forest Service wanted to avoid this as well and created a management plan for the region that it hoped would protect enough of the bird's habitat to keep it off of the list. Nevertheless, this middle-of-the-road plan had no friends. The timber industry believed it was too restrictive while environmentalists argued the plan would still lead to the owl's extinction.[18]

The issue simmered throughout much of the 1980s as the timber industry lobbied to keep the bird from being listed, environmentalists tried to maximize protections, and the Forest Service tried to meet every party's needs. By the end of the decade, however, a campaign to list the bird as threatened gained substantial momentum, escalating tensions from a simmer to a full boil. As a number of environmental organizations filed lawsuits to force the Fish and Wildlife Service to list the bird, advocates on both sides were taking to the streets. At a rally in Olympia, Washington, scuffles and shouting matches broke out between loggers and environmentalists. Throughout Oregon, bars hung spotted owls in effigy, and bumper stickers that read "I Love

Spotted Owls ... Fried" proliferated. Perhaps most representative of how hysterical the issue became, Smokey Bear and the Forest Service's other mascot, Woodsey the Owl, received death threats that prevented them from making their annual appearances in Portland's Rose Festival Parade.[19] Throughout this period, the USFWS was hesitant to list the owl because, unlike with the snail darter, it understood the political implications and were much less willing to base its decision purely on science even though the ESA required it to do so.

Many of Oregon and Washington's political representatives continued to lobby on the timber industry's behalf, but after the Fish and Wildlife Service listed the owl in June 1990 and designated eleven million acres in the two states as critical habitat a year later, the timber industry's case had few avenues of recourse. After a federal court upheld these decisions, the timber industry was forced to accept this new reality.[20] The older model of conservation had protected forests for the benefit of extractive industries like logging, but under the purview of environmentalism, forests were maintained for the benefit of wildlife like the spotted owl. This shift did not sit well with many in the Old West who believed the environmental movement had gone too far and gave credence to criticisms made by the Sagebrush Rebellion and its later iterations.

In many ways, the spotted owl controversy was the penultimate debate that had been brewing over the previous decades. As purely as possible, it pit the values of economic development against environmental protection because it did not deal solely with a single project or small region, but with a vast area, almost as large as New Hampshire and Vermont combined. It was a debate between conservationism and environmentalism, and although environmentalism and the Endangered Species Act fared better than they did at Tellico, the episode also united and enraged opponents for years to come.[21]

Most significantly, the episode helped complete the transformation of the fundamental nature of politics in the interior American West in a way that would resonate for decades and directly influence how the debate over the Bitterroot grizzlies would unfold. States like

Montana and Idaho had adhered to a progressive brand of politics ever since the rise of populism and William Jennings Bryan in the 1890s. When Eugene V. Debs ran for president as the candidate for the Socialist Party, from 1900 through 1912, western states like Montana and Nevada consistently voted for him at higher rates than did others. The interior West became a stronghold of organized labor where loggers and miners otherwise would have been unprotected from the large companies for which they worked. In the 1930s, the ascendance of Franklin Delano Roosevelt strengthened these loyalties as farmers and unskilled laborers benefitted from the New Deal's liberal policies. These trends persisted into the 1970s when Montana enacted one of the nation's most progressive state constitutions in 1972 and elected liberal leaders like Mike Mansfield and Lee Metcalf, who served from 1953 to 1978.[22] However, by the end the decade, the West, and especially the Northern Rockies, had joined the South as one of the most conservative regions of the country.

Because extractive industries had been the only powerbrokers in the West since settlers first migrated to the region, these industries had been willing to support progressive policies that supported the growth of their industries, but as the environmental movement began to challenge their hegemony, they suddenly had to fight to hold onto what they already had. This wholesale shift led to a more conservative outlook that reached beyond the purview of policies immediately affecting farming, ranching, logging, and mining and influenced how the West's old powerbrokers addressed every potential change. By the 1980s this shift was complete, and although the debate over the spotted owl did not directly affect Idaho and Montana, extractive land users took valuable lessons from the ongoing saga that confirmed their adherence to conservative politics. The snail darter episode had drawn attention to endangered species issues on a national scale, but many of the issues that the spotted owl controversy raised were uniquely western. Throughout the attempt to bring grizzlies back to the Bitterroots, opponents would hold up the spotted owl controversy as a cautionary tale and it became a rallying call for the opposition.[23]

While the majority of criticism for the ESA came from conservatives who did not agree with the act's principles, by the early 1990s, many environmentalists did not have a much better opinion of it either, believing it had not produced enough significant achievements. Although the act had saved a few species from extinction, some were relegated to breeding in captivity, and commentators questioned whether the act was even capable of enabling full recovery. Critics also pointed out that the act's few successes dealt solely with charismatic megafauna and had been unable to recover less-prominent species. Other critics believed the act was ineffective because it was not designed to protect full ecosystems. Although section 4 of the act stipulated that listing and delisting decisions were to be based solely on science, many doubted how strictly the government adhered to this provision. Finally, one commentator asserted that the act was in trouble not only because it ignored private interests and lacked proper structure, but "because it challenges, for many Americans, their beliefs about themselves."[24] Compared to many criticisms of the ESA, this comment cut to the heart of the real issue. The challenges facing the act lay less in the shortcomings of its legal structure than in its transformative social and cultural implications. The act forced Americans to put wildlife ahead of themselves, and by the late 1980s and early 1990s, fewer Americans were willing to make that sacrifice than had been the case twenty years earlier.

At the verge of the 1990s, some environmentalists began to recognize this new reality, while many others continued to press their agendas without seriously considering that times had changed and environmental protections could mean serious economic sacrifices for certain Americans. Additionally, similar to the Civil Rights Movement in the late 1960s, a number of environmental groups, such as Earth First! and Greenpeace, radicalized and turned to violence. As a result, the environmental movement found itself at a crossroads, plagued by internal conflict, unsure of how or where to proceed. In the 1960s and early 1970s, environmental organizations rode a wave of popular support and even though their budgets were small, they were inspired and passionate, and they had momentum from fighting a moral battle against faceless bureaucrats and corporations that thought nothing of degrading the environment. Their staffs, which often went unpaid, were fiercely committed to their causes, and this zeal was largely the source of their success.[25] By the 1980s, however, the fundamental makeup of environmental activism began to change.

While the rise of the Reagan administration and the appointment of James Watt undoubtedly dealt a blow to the movement's ability to pass legislation and institute reforms, its leaders were able to turn the administration's hostile stance toward the environment to their advantage by using it as a way to create urgency and increase support. Reagan and his polices became the common enemy that energized and united the movement. For mainstream environmental organizations, the arrival of Reagan and Watt on the national stage caused their memberships to skyrocket, and in turn, so too did their budgets. As a result, prominent organizations like the Sierra Club, Audubon Society, National Wildlife Federation, and Wilderness Society became massive lobbying groups run by business administrators focused on finances and meeting a bottom line. With budgets in the tens of millions, the environmental movement became just another lobby, no different from the large corporate interests against which it had initially fought. Because of this, the movement lost much of its soul.[26]

In this same vein, much of the public was growing frustrated with the tactics of the mainstream environmental organizations. As they grew in size, these groups were turning away from the grassroots tactics that had originally characterized the movement. Instead of consulting local people or state governments, environmentalists engaged their national support networks to push their agendas. And when that did not work, they were quick to rely on the courts to enforce their will. Because these decisions took power away from states and local communities, many Americans viewed this trend as undemocratic, eliciting the charge that the environmental movement was out of touch with the majority of the country. Furthermore, a divide between national organizations and local organizations grew as national organizations became more willing to compromise on particular issues with the idea that no one battle was worth risking their long-term influence whereas local organizations that focused on just a single or small number of issues were less willing to compromise and more dedicated to environmentally pure solutions.[27]

On the other hand, for groups like Earth First! and Greenpeace, the conservative policies of President Reagan and Secretary Watt pushed them further to the left and inspired them to resort to tactics that bordered on ecoterrorism—spiking trees and attacking shipping vessels. While this benefitted the environmental movement to some degree—the more radical groups making the mainstream organizations appear less extreme—a certain segment of the environmental movement's opponents drew no distinctions between the Sierra Club and the more radical factions of the movement. For people who believed that groups like the Sierra Club had gone too far, Earth First! was absolutely frightening and further convinced them that the environmental movement was out of control.[28]

Throughout the 1980s, the environmental movement polarized the public—supporters became stauncher and opponents more dogmatic—and when the Clinton administration assumed power, the movement was set for another change. With an administration perceived as friendly to the goals of environmentalism, many Americans lost the sense of urgency they had felt under Reagan, and memberships in the big environmental organizations dropped precipitously. Many of the large organizations were forced to downsize as a result, and the smaller, local organizations rose in prominence once again. Consequently, many keystone environmental groups became more fundamentalist and extreme in their demands as a way of maintaining their relevance and ideological connections to the smaller groups. Nevertheless, the country's economic situation remained uncertain, which made many Americans less likely to make the sacrifices necessary to protect animals such as the spotted owl and fueled criticisms that environmentalists were out of touch. In reaction to these sentiments, a small cadre of environmentalists was willing to adapt, but they were the exception.[29] As a result of this break within the movement, Bitterroot grizzly bear advocates would end up fighting with each other as much as with their Wise Use counterparts.

While the country was locked in an ideological struggle that pit burgeoning conceptions of a more sensitive relationship with the natural world against deeply entrenched values that prized economic development, grizzly bear recovery remained under the radar, moving forward slowly, but steadily. At first, however, despite the Endangered Species Act's legislative protections, serious, proactive efforts to promote recovery did not begin for almost another decade, and grizzly bear populations continued to decline. In the early 1980s, the USFWS hired a coordinator to take control of the recovery program, and in early 1982, the USFWS released its first recovery plan. The Grizzly Bear Recovery Plan (GBRP) was essentially a blueprint for achieving recovery that laid out the goals and methods needed to restore grizzly bears and their habitat to sustainable levels. According to the GBRP, the actions needed to grow and stabilize the bear's population included minimizing human-bear conflict, limiting habitat loss or degradation, improving habitat, understanding better the relationship between population density and habitat, developing techniques to move bears into regions with low populations, improving public relations through education, and continuing research.[30]

Still, building a healthy population of bears depended on the cooperation of a number of competing agencies including the Forest Service, Park Service, and Fish and Wildlife Service, along with the relevant state wildlife agencies. To address the difficulties these agencies initially had working together, Rocky Mountain Regional Forester Craig Rupp helped organize the Interagency Grizzly Bear Committee (IGBC) in 1983. The IGBC included members from the state wildlife agencies from Idaho, Montana, and Wyoming, four federal agencies, local and tribal representatives, and multiple nonvoting scientific advisers. Although the group dedicated most of its attention to recovering the bear in its strongholds, preliminary efforts to recovery grizzlies in the Bitterroots commenced by the middle of the decade.[31]

Because the USFWS listed the Bitterroot ecosystem as an evaluation area, the agency needed to conduct further research to determine the status of the region's grizzly population, the quality of its habitat, and its suitability as a recovery area. The same was true for the North Cascade and San Juan ecosystems. Through the early 1990s, efforts in the North Cascades and Bitterroots fell in step with each other, but the USFWS quickly decided not to pursue recovery in the San Juan ecosystem because it was so far away from the other recovery areas.[32]

The first step needed was to establish the status of the region's resident grizzly population. In 1985 Wayne Melquist conducted the first study in the Bitterroots designed to estimate the size of the region's grizzly bear population. This study focused on the Clearwater National Forest in the northern Bitterroots, where a population of grizzlies would most likely reside, and collected and analyzed the eighty-eight grizzly bear sightings reported from 1900 to 1984 in addition to doing ground and aerial reconnaissance. Although eighty-eight grizzly sightings were reported, Melquist knew there was a high probability that many were actually of black bears. To the untrained observer, the two species are easily confused as black bears are often a cinnamon color, similar to grizzlies, so Melquist's challenge was to verify these reports.

On flyover and ground surveys, Melquist saw black bears and their sign, but no grizzlies. For the ground surveys, he hiked more than 215 miles and found many tracks, scat, and tree rubs, but he could not link any of these to grizzlies. For the other part of his study, Melquist subdivided the eighty-eight sightings into four categories—confirmed, probable, highly possible, and possible but lacking sufficient information—and according to his system, only fourteen of the seventy-four sightings that occurred since 1960 ranked as highly possible. From this data, Melquist concluded that a "few grizzly bears probably occupy, at least temporarily, portions of the Clearwater

National Forest and adjacent areas." However, he also noted that "there is considerable evidence that a viable population no longer exists in the Selway-Bitterroot Wilderness."[33]

That same year, Bart Butterfield and John Almack led a similar study to determine the quality of the habit in the Selway-Bitterroot Wilderness and whether the region could support a population of grizzlies. The detailed survey—which analyzed the wilderness area for the quality of its denning sites, cover, and availability of seasonal food sources—concluded that the region was not only home to many diverse habitat types, but contained the seven essential characteristics of grizzly bear habitat described by John Craighead. From this conclusion, Butterfield and Almack determined that the region was well suited for hosting a population of grizzly bears. Their study was restricted to the wilderness area, but the two researchers went a step further and suggested that the entire ecosystem, taken as a whole, elevated the region to a ranking of "ecologically superior" for grizzlies.[34] While Melquist's findings were necessary to establish the status of the population, Butterfield and Almack's findings were ultimately more significant because they meant that recovery would be possible. If the proper habitat did not exist, it would not have mattered if a few remnant bears roamed the area because they would not be able to establish a sustainable population.

Because Melquist's study had focused solely on the Clearwater National Forest, another researcher, Craig Groves, expanded Melquist's work two years later to include the entire Bitterroot ecosystem. Just as Melquist had done, Groves compiled and analyzed all of the reported sightings from the region over the previous century. Groves determined that the Nez Perce National Forest, south of the Clearwater, had only produced three sightings between 1979 and 1985, and that the last sighting in the Salmon National Forest, even further south, occurred in 1977. As a result, he too concluded that a few bears potentially occupied the region temporarily or seasonally, but no true population existed.[35]

Melquist and Groves's surveys mostly focused on reported sightings by third parties and did not feature much on-the-ground fieldwork. To fill this gap, in 1991 a team of scientists from Idaho's Department of Fish and Game conducted a remote-camera study in the northern Bitterroots in hopes of documenting grizzly bears in the ecosystem. Of the 825 photographs that their motion-sensor cameras captured, 296 of them were of wildlife. Black bears were the most photographed animal followed by deer and elk, but no grizzly bears appeared even though the team placed the cameras in the areas where Melquist and Groves had identified "highly possible" sightings.[36] While many people wanted to believe that grizzly bears still roamed the area, the evidence was mounting against them.

That same year, another study, which took five years to complete, confirmed Butterfield and Almack's findings. It concluded that despite the lack of evidence of any grizzly bears, the Bitterroot ecosystem contained the necessary space, isolation, denning opportunities, vegetation, and food to support a viable population. Grizzly bears eat meat, herbaceous vegetation, and fruit and nuts, and the study determined that the Bitterroot ecosystem was rife with elk, deer, sedges, rushes, blue bunch wheatgrass, whitebark pine nuts, huckleberries, snowberries, honeysuckle along with many other favored grizzly bear forage. A third team of biologists, led by the USFWS's grizzly bear recovery coordinator Christopher Servheen, reported to the IGBC at its annual December meeting in Denver that the habitat in the Bitterroots was ample enough to support a population of two hundred to four hundred bears, which would mean roughly only one bear per 26.4 square miles. This was approximately equivalent to the density of the population in the Northern Continental Divide ecosystem, which encompassed Glacier National Park, and supported one bear for every fourteen to twenty-two square miles.

The IGBC accepted these conclusions, and at the end of 1991, it elevated the status of the Bitterroot ecosystem to that of a full recovery area. The IGBC also authorized the compilation of a Bitterroot chapter for the recovery plan, which would spell out what would be needed to bring a sustainable populations of grizzlies back to that

region.[37] Although these studies confirmed that grizzly bear reintroduction, from a biological basis, looked promising, they all acknowledged that the human element was the biggest unknown factor and
would determine whether recovery succeeded or not. The political
battles that would embroil the recovery of grizzlies in the Bitterroots
were years away, but these premonitions would prove more accurate
than their authors could imagine.

Despite the fact that environmentalism had replaced conservation as the dominant environmental ethic during the 1960s and
1970s, many Americans, Old Westerners especially, remained loyal
to conservation and rejected environmentalism's moral dimension. As a stagnant economy and disillusionment with big government took hold by the end of the 1970s, Old Westerners seized
their opportunity and attempted to steer the country back to these
Pinchot-era policies, first in the form of the Sagebrush Rebellion
and later as the Wise Use Movement. While this was unfolding,
the environmental movement was undergoing an internal crisis.
It had grown considerably over the previous decades, transforming from a small grassroots movement to a national lobby with
millions of dollars at its disposal. In the process, it had lost some
of its appeal. Many in the movement were trying reinvigorate it,
but by the beginning of the 1990s, their success had been limited
and their direction unclear.

   This would be the environment through which Bitterroot grizzly
recovery would have to navigate. Nearly a century after Gifford Pinchot and Theodore Roosevelt introduced conservation to the American lexicon and made sustainable natural resource management
the mission of the federal government, Americans' environmental
consciences were more in flux than ever. Having Smokey and Yogi
as cartoon symbols that depicted bears as amicable pals was one
thing, actually returning grizzlies to a region where people worked,
lived, and recreated was entirely different. Bear advocates would
quickly discover that as significant as bears' public image makeover

had been, Bitterroot recovery would be no easy task. But before it reached the national stage, the proposal to reintroduce gray wolves to central Idaho and Yellowstone pitted Old West conservationism against environmentalism and the New West in a way that would shape the debate over Bitterroot grizzly recovery at every step.

3

Wolf Recovery Sets the Stage

The big day had finally arrived. Over the previous week, a cadre of government biologists had been in northern Alberta capturing wolves to be reintroduced to central Idaho and Yellowstone National Park. They had planned to release fifteen wolves at each site that winter, but in this first round, they captured only eight wolves that would be sent to Yellowstone and four that were destined for Idaho. The wolves were packed into aluminum crates for their journey south, and as they made their way through Paradise Valley, north of Yellowstone, a caravan of park rangers escorted them, ready to protect the animals from any overzealous Old West vigilantes who wanted to harm the park's newest residents.[1]

The caravan drove through the small gateway town of Gardiner, Montana, at the park's northern entrance, and as it entered the park, the trailer pulling the wolves passed under the Roosevelt Arch. The arch, which serves as the park's official entrance, is a massive basalt structure, named for the twenty-sixth president, who gave a speech at the partially constructed entryway while touring the park in 1903. Prominently stamped at top of the arch are the words "For the Benefit and Enjoyment of the People," which was the language used in the law establishing Yellowstone in 1872. These words seemed fitting as hundreds of people lined the sides of the road on the cold January morning in 1995, watching as the wolves returned to the park after a seventy-year absence. As the caravan passed by, the crowd cheered. Some supporters let out enthusiastic howls, while others tried to hold back tears. News teams from all around the world had arrived to document the event. One reporter proudly claimed, "Being here is my reward for spending two and a half years in Bosnia." Photographers clamored for pictures of the park's newest four-footed celebrities, and the event became so frenzied that a report in *High Country News* the following month focused solely on the "circus atmosphere" that pervaded coverage of the release. Over the next few weeks, the Park Service even had to establish a lottery in order regulate the number of journalists allowed to view the wolves in the one-acre holding pens that were their temporary homes before being released fully into the park.[2]

The day was a historic one, the culmination of a twenty-year fight, and many people in attendance could sense its significance. The director of the Fish and Wildlife Service, Mollie Beattie, was on hand as was Secretary of the Interior Bruce Babbitt. After passing through the arch, the caravan stopped briefly and Babbitt stepped down from his vehicle to ponder the deeper meaning behind the event. With dozens of television cameras eagerly awaiting his thoughts, Babbitt triumphantly declared, "This is a day of redemption and a day of hope. It's a day when the limits of what is possible have been greatly expanded because we are showing our children that restoration is possible, that we can restore a community to its natural state." The redemption to which Babbitt referred was a direct allusion to the wrongs committed by Old Westerners who had killed off wolves earlier in the century. With Babbitt's words ringing in their ears, New Westerners believed wolf reintroduction to be not just a victory for ecological balance and a more sensitive understanding of the natural world, but a sign that their biocentric approach to natural resource management had trumped the anthropocentric model espoused by their Old West counterparts as the dominant ethic governing the region.[3]

However, the pomp and circumstance taking place at the arch was just part of the story. After the caravan took off again, it headed into

Yellowstone's Lamar Valley where the wolves' crates were loaded onto sleds and hauled a mile into the backcountry to the holding pens where they would acclimate to their new homes over the next few weeks before being released fully into the park. Babbitt, Beattie, and a handful of biologists lugged the crates the final few yards to the holding pens, but after all the crates were securely inside, they closed the gates without letting the wolves out of the claustrophobic entrapments they had endured for the previous thirty-plus hours. Despite all the festivities that had unfolded earlier that day, much of it had been just for show. That morning, the Mountain States Legal Foundation had filed a last-minute appeal to delay the wolves' release, and a federal appeals court granted the stay. It was an underhanded move because the wolves were already in transit, and they would not be able to survive long in their crates, which were designed only for transportation. Many of the wolves were starting to falter and park biologists feared the worst if the animals were forced to remain in their cages much longer. Luckily for the wolves, the court lifted the stay late that evening and park biologists were able to release them into the pens.[4] Ultimately, the day was a victory for the wolves, and Babbitt's triumphalist pronouncements withstood the eleventh-hour legal chicanery, but the day's events also indicated that the New West's victory was not as far-reaching as some wanted to believe. They had won the battle, but the war was far from over.

From 1975 to 1995, gray wolf recovery in the Northern Rockies was a defining issue whose circuitous path toward victory reflected many of the larger themes that influenced the environmental movement at large. When the Fish and Wildlife Service formed the Northern Rocky Mountain Wolf Recovery Team in the mid-1970s, environmentalism still held its mandate. However, by the time the team produced its first recovery plan in 1980, the movement had lost much of its support, which prevented the USFWS from approving a plan until 1987. Pressure from conservatives delayed Congress from funding an environmental impact statement for another few years, but as the environmental movement turned more radical in the post-Reagan era, hard-line environmentalists pushed back against the relatively moderate plan the government backed. Victory for gray wolf recovery was a long, slow march, and as the events of that cold January morning in 1995 demonstrated, the likelihood of easing tensions between the Old West and the New West was minimal.

While wolf recovery provides an exemplary case study with which to examine the trajectory of the environmental movement over this period, more significantly, for our purposes, it can be seen as a precursor that fundamentally shaped the debate that followed over Bitterroot grizzly recovery. It brought the divide between the New West and Old West to the forefront of regional politics and laid the groundwork for the debate over Bitterroot bears. Not only were many of the issues in the two controversies similar, but many of the same politicians and activists who were engaged in wolf reintroduction were just as involved in the deliberations over grizzly bears. Without a doubt, Bitterroot grizzly recovery could not have unfolded the way it did if wolf recovery had not preceded it, thus making a thorough understanding of the issues surrounding wolves paramount.

In many ways, the history of wolves in the West followed a similar trajectory to that of grizzlies. At first, because wolves did not pose the same threat to humans that grizzlies did, early travelers, such as trappers, mountain men, and westward bound settlers, largely ignored wolves and started hunting them for their pelts only after populations of more prized animals had been decimated. But as ranching became more prevalent in the 1870s and 1880s, and the region as a whole became more settled, ranchers, in particular, developed a virulent hatred for wolves. Although settlers may have disliked grizzlies because they threatened human supremacy, bears scavenged for much of their food. Grizzlies would sometimes kill livestock, especially sheep, but they did not kill with the same consistency as did wolves. Wolves hunted for the large majority of their sustenance, so their presence posed a more direct threat to ranchers' livelihoods than did bears. Wolves had adapted to hunting wild game animals, but as those populations decreased, wolves lost their traditional food sources and turned to the cattle and sheep that had replaced them. Because of this, ranchers began killing wolves out of economic self-interest. This purely practical need, however, eventually bred a culture of hatred that ingrained itself among western ranchers and land owners and went beyond economic necessity, dictating absolute, wholesale annihilation as its goal.[5]

Predation of stock increased along with ranching, and as it did, so did ranchers' hatred for wolves. By the 1880s, ranchers across the West started offering bounties on wolves. Because these efforts were disorganized, bounties proved to be an ineffective way of eradicating the hated predator at the population level. Nevertheless, they were wildly popular and led to the deaths of thousands of wolves. To coordinate their campaign more effectively, ranchers began forming stockmen's associations, which took charge of wolf reduction programs and lobbied for state and federal programs to aid in their campaign. Eventually, the U.S. Biological Survey, a predecessor of the Fish and Wildlife Service, took control of the program and proved itself much more efficient and effective at this task. Most regularly, its agents planted carcasses laced with one of a number of poisons, most often strychnine, so that wolves would unwittingly eat the poison and drop dead a few feet from their last meal. Because conservation's utilitarian umbrella did not include wolves, early conservationists and advocates for the humane treatment of animals condoned these killings. Some people objected to this wholesale destruction, but newspapers commonly depicted wolves as bandits, criminals, and desperadoes, and eradication efforts were immensely popular. They were so successful that by the 1920s, wolves were extinct across much of the West, leaving only small populations in Minnesota and Michigan.[6]

Similar to that of grizzly bears, the public image of wolves received a makeover, national in scope, in the post–World War II era that went hand in hand with the rise of the environmental movement. Researchers began learning more about the animal, books celebrating wolves reached mass audiences, and nature shows on television shed a positive light on the once-hated animal. People began to relate to and celebrate the positive traits that wolves shared with humans such as mating for life, caring for young, and killing only out of necessity. As much as any other animal, wolves came to embody a freedom and wildness unsullied by civilization's advances that many environmentalists embraced. New Westerners brought this enhanced veneration for the wolf to the Northern Rockies, where the presence of ranching would have otherwise kept the animal's popularity at a minimum. Much like that of the grizzly bear, the rise of the environmental movement and the remaking of wolves' public image went hand in hand with the rise of the environmental movement, so that by the time Congress passed the Endangered Species Act, the national reputation of wolves had been on the rise for a number of years. In fact, their public image had undergone such a wholesale reversal that the Fish and Wildlife Service listed the gray wolf as endangered across the entire lower forty-eight states in 1973 just as the Endangered Species Act became law.[7]

The following year, while the environmental movement still held its clout, the USFWS established the Northern Rocky Mountain Wolf Recovery Team, an interagency group composed of representatives from federal and state agencies that had a stake in wolf recovery, to formulate a recovery plan for gray wolves. Members of the recovery team knew that wolf reintroduction would be controversial, but in retrospect, they were still naïve. They simply had no idea of the extent to which wolf recovery would become a divisive issue that would gain a national audience and rage for nearly twenty years. Furthermore, because the Endangered Species Act was so new, the group was charting untested waters and had no idea how closely people on both sides of the issue would scrutinize the plan's every detail. The plan they crafted was vague. Federal regulations required the plan to establish goals for recovery and the areas in which recovery would be a priority, but this first plan did not delineate the specific areas where wolves would be recovered, how to facilitate recovery, or how to deal with wolves that preyed on stock. Furthermore, it neglected to include Yellowstone as a recovery area even though many environmentalists considered it ideal. Complicating matters even more, the team did not release the plan until 1980, after the Sagebrush Rebellion had commenced. By that point, the snail darter episode had brought increased scrutiny to endangered species issues, and after the spotted owl became a source of tension between the Old West and the New West a few years later, any new endangered species programs in the region were greeted with suspicion and controversy. Wolf recovery would have been a contentious regional issue in any case, but these other controversies elevated its prominence to a national level. With this tumultuous political climate in mind, the plan's timeline did not initiate recovery steps until 1987.[8]

Because the Reagan administration was in office, the environmental movement's ability to effect change was nearing an all-time low, and the USFWS was unwilling to approve the plan. Two years later, however, the environmental movement had regained some of its status, using President Reagan's sympathy for the Sagebrush Rebellion to rally support. This, in part, led the USFWS to revamp the recovery team, appointing Bart O'Gara as team leader. O'Gara was known as the type of person whose voice could resonate with both ranchers and wolf advocates. John Weaver, a Wyoming biologist who had studied wolves in Yellowstone for the Park Service, also joined the team. Weaver was a passionate advocate for wolves and became an able adversary to Joe Helle who represented the Montana Wool Growers Association and defiantly opposed wolf recovery under any condition. Helle was not a member of the recovery team, but he attended every meeting and made his voice heard—as did Hank Fischer, a Missoula-based conservationist who had represented Defenders of Wildlife since 1977. The recovery team's meetings often bordered on hostile as Helle took a "hell no" approach and refused to support wolf recovery efforts on any terms. Despite his intransigence, the recovery team made progress, demarcating three recovery areas: the Northern Continental Divide, Greater Yellowstone, and Bitterroot ecosystems—the last of which included the Frank Church–River of No Return Wilderness Area in central Idaho.[9]

By 1982 wolf recovery had progressed steadily, remaining out of the public eye, but when Weaver drafted a set of wolf management guidelines for the recovery team to consider, Helle sounded the alarm. By convincing politicians, ranchers, extractive land users, and other Old Westerners that these guidelines would bring restrictions on logging, mining, and grazing across the West, Helle touched off a firestorm. The recovery team was bombarded with angry letters, and Idaho politicians accused the recovery team of having reintroduced wolves secretly. One of Idaho's senators, Larry Craig, held hearings in Boise and Grangeville to address concerns about wolf recovery and make it clear to anyone wishing to reintroduce wolves just how unpopular the notion was. On top of this, because Ronald Reagan and James Watt were decidedly antiwolf, recovery efforts stalled again.[10]

In response to criticisms from Old West constituents like Joe Helle and other conservationists who feared the land-use restrictions that could accompany programs like wolf recovery, the federal government was taking steps to mitigate those restrictions. In 1982 Congress passed an amendment to the Endangered Species Act that gave the law greater flexibility in order to ensure that endangered species restoration and extractive land use were not mutually exclusive. The first part of the amendment actually strengthened the law by clarifying that the federal government should make listing decisions *solely* on the basis of science without taking into account the economic or social ramifications that listing might bring. However, in seeking to accommodate economic development and extractive land users, the amendment also added language to section 10 that allowed the USFWS to introduce "non-essential, experimental" populations of endangered or threatened species in areas where no resident populations existed. Most notably, this relaxed the stipulations of section 7 that protected a listed species' habitat. Because nonessential experimental populations were not considered vital to the species' continued existence, their habitat could not be protected using the "critical" status. As a result, Helle and other land users would be able to ranch, mine, harvest trees, and farm on land even if they shared the land with an experimental species. This flexibility also meant that individual animals could be killed under certain circumstances. Despite these stipulations, the law maintained that all policy-level decisions would have to lead ultimately to the species' recovery. So although a marauding wolf or bear that had developed a taste for livestock could be destroyed, the amendment dictated that policy-level decisions must encourage the growth of the species' overall population.[11] The amendment was significant not only because it helped make wolf restoration a more realistic possibility, but because it laid the ground work for the more inclusive, collaborative approach to endangered species restoration that the architects of Bitterroot grizzly recovery would capitalize on a decade later.

The negative attention that Helle brought to wolf recovery had forced the Fish and Wildlife Service to reject the 1980 recovery plan. So in 1983 the team went back to the drawing board, and after much wrangling, it produced a new recovery plan that was much more detailed and specific. According to the plan, for wolves to be recovered and taken off the endangered species list, at least ten breeding pairs would need to inhabit each of three designated recovery areas for at least three consecutive years. Additionally, the team decided that while natural recovery through migration from Canada would be suitable for central Idaho and northwestern Montana, recovery in Yellowstone would require human-assisted reintroduction. Finally, taking advantage of the 1982 ESA amendment, the plan stipulated that the reintroduced population would be considered experimental.[12] Although environmentalists hoped the plan's greater level of detail would mollify Old Westerners who suspected that wolf recovery would be dictated by the whims of environmentalists and be subject to little or no accountability, the team had developed it without much input from these Old West opponents. Because most Old Westerners had taken a "hell no" approach to the subject, New Westerners felt little compunction to incorporate their feedback. While this strategy may have been the product of circumstances, the plan continued to be as unpopular as ever among Old West politicians and their constituencies, thus ensuring that wolf recovery remained fiercely controversial.

If the experimental designation was supposed to allay ranchers' fears, no one told them; and even though the recovery team's new plan was an important step toward restoration, continued opposition from Old Westerners prevented the Fish and Wildlife Service from approving it. Over the middle years of the decade, wolf supporters held meetings throughout Idaho, Montana, and Wyoming trying to convince farmers and ranchers of wolf reintroduction's merits. Despite assurances that wolf restoration would not bring weighty federal restrictions on land use or lead to significant losses in livestock, most ranchers and their political representatives remained adamant in their opposition. "I think Montana needs wolves like it needs another drought," charged Montana's Republican congressional representative, Ron Marlenee. He surmised that wolf recovery was "just another ploy by 'green bigots' to block economic development of Montana." Republican senator Alan Simpson added, "In a state like Wyoming, where we rely heavily on our livestock industry, the wolf is not actually a very welcome new resident." Simpson later added, "Wolves eat things—human and alive," a common misconception that wolf opponents frequently recited. Others argued against wolf reintroduction because of its high cost or, more ironically, the fact that it could potentially interfere with grizzly bear recovery in the Yellowstone ecosystem, a project that few opponents of wolf recovery were enthusiastic about either. Perhaps more adeptly, some wolf supporters suggested that many Old Westerners who opposed reintroducing wolves had fathers and grandfathers who helped eradicate them earlier in the century, and reintroducing them would be admitting that their predecessors had been wrong to kill them in the first place.[13]

While wolves' opponents argued against their reintroduction on economic grounds, wolf supporters used ecological as well as a legal arguments for their recovery. The law creating Yellowstone as the first national park in 1872 discussed protection against the unwonted destruction of game species. For many years, park managers interpreted this to include killing predators that preyed on more "desirable" species. This meant targeting not only wolves and coyotes, but species such as white pelicans that fed on cutthroat trout. In 1926 rangers killed the last wolf in Yellowstone, and for a few years the park thought it had satisfied its mandate. However, within a few decades, biologists began to realize that the wolves' absence left a hole in the ecosystem. Few other predators existed to cull ungulate herds of their sick and old, and elk, bison, and deer populations skyrocketed as a result. The absence of wolves created a ripple effect that negatively impacted species as big as grizzly bears and as small as beavers and had detrimental effects on the growth of certain types of vegetation as well. As a result, many people who supported wolf reintroduction did so out of a desire to restore balance to the ecosystem. Yellowstone's superintendent made it clear when he affirmed, "Biologically, the case is clear . . . the wolf should be here."[14]

The strongest arguments on behalf of wolf restoration were ecological in nature, but many people supported wolf recovery for the same emotional and idealistic reasons that had inspired the environmental movement. For these people, wolves embodied romantic notions of wildness and freedom that was celebrated by environmentalists and New Westerners alike. Reintroduction symbolized a more deferential relationship with nature reflective of the ethos espoused by Aldo Leopold that "a thing is right when it tends to preserve the integrity, stability, and beauty of the biotic community. It is wrong when it tends otherwise." And as far as Yellowstone recovery was concerned, many simply wanted to improve the park experience for visitors.[15]

Wolf advocates knew the fastest path toward recovery required winning political support, but few of the region's politicians possessed the magnanimous nature or political capital to weather the inevitable push back that would accompany supporting reintroduction. Max Baucus had been Montana's Democratic senator since 1979 and gen-

erally was a friend of the endangered species program and environmental reform. However, his family owned one of the largest sheep
ranches in the state, and he refused even to discuss the wolf issue.
Pat Williams, Montana's Democratic representative in the House,
whose interests generally aligned with those of the New West, had
one of the strongest environmental records of anyone in Montana,
Wyoming, and Idaho, and he had spent many years lobbying for the
creation of more wilderness areas. Although he had no ideological
aversion to wolves, he understood the power of the Old West and
was savvy enough to realize that supporting wolves could threaten
his political career. Another possible candidate was Idaho's governor, Cecil Andrus. Andrus had been secretary of the interior under
President Carter and had displayed an environmentally progressive
streak while in office. His crowning achievement as secretary was
the Alaska Lands Act, which protected more than 104 million acres
and created twelve parks and 56 million acres of wilderness. Once
back in Idaho as governor, however, his priorities changed. The Old
West reigned supreme in Idaho, making little room any for progressive environmental reforms. Like Williams, Andrus understood this
political reality, and he outright opposed wolf reintroduction.[16]

Because wolf reintroduction would have the greatest direct impact
on Idaho, Montana, and Wyoming, and the political delegations of
those states demonstrated no willingness to support it, other congressional and bureaucratic leaders were also hesitant to support it.
In the early 1980s, few federal agencies even wanted to discuss the
issue. A few years later, agencies were willing to discuss it, but most
of them were far from supportive. Frank Dunkle, director of the U.S.
Fish and Wildlife Service, initially was adamant about blocking wolf
recovery from proceeding and refused to approve the wolf recovery plan. While speaking to the Wyoming Woolgrowers Association,
Dunkle pointed to his neckwear and announced that "the only wolves
that I will bring to Wyoming . . . are on this tie." Furthermore, Dunkle
promised to use every bureaucratic means at his disposal to delay
wolf reintroduction.[17]

Once political and bureaucratic support proved out of reach, wolf supporters decided the best way to move reintroduction forward was through their national support networks. Instead of trying to modify their approach to wolf recovery and directly address the concerns of their opponents, wolf advocates turned to one of the environmental movement's battle-proven strategies and looked for ways to overwhelm local opposition with a national publicity campaign. In 1985 Defenders of Wildlife sponsored a survey in Yellowstone, which determined that 81 percent of visitors agreed that wolves should have a place in the park and found that 60 percent confirmed that if wolves could not return to Yellowstone on their own, they should be reintroduced. That same year, Defenders of Wildlife commissioned a museum exhibit called *Wolves and Humans*, which toured the West, making stops in both Boise and Yellowstone. The award-winning exhibit focused on human perceptions of wolves and challenged visitors to appreciate the common bonds that wolves and humans share. More than two hundred thousand people viewed the exhibit in Yellowstone alone and support for wolves flourished. Additionally, Reagan's new Park Service director, William Penn Mott Jr., displayed an unbridled enthusiasm for wolf recovery unlike that of any other bureaucrat at the federal level. Mott stressed the need for public education and argued that to reintroduce wolves, advocates needed to win the public relations battle.[18]

Because the wolf was listed as endangered, the Endangered Species Act required that it be recovered, and Hank Fischer recalled that early in his career he "thought the law alone was enough to save wildlife." However, he quickly learned otherwise, and with Mott's advice and precedents from previous endangered species battles, he and other wolf advocates strengthened their media campaigns. In addition to the museum exhibit, which made stops in New York, Washington DC, and Boston, Defenders of Wildlife sponsored public meetings, school assembly programs, lecture series, and distributed brochures throughout various towns in the Northern Rockies. Even though the congressional delegations of Montana, Idaho, and Wyoming and their extractive land-using constituents remained hostile toward the program, national media outlets showered wolf recovery with positive press; and what had been a regional issue assumed national dimensions. As a result, national support thrived, and Hank Fischer would later identify the public media campaign that began with the *Wolves and Humans* exhibit as a turning point in the battle.[19] This was essentially the model of 1960s and 1970s environmentalism. Instead of sitting down with their Old West opponents and trying to work out a solution, wolf advocates summoned their national support and media networks. By some measures, this strategy was wildly successful, but it further angered Old Westerners and made them feel marginalized in the process. Fischer, as well as others, believed this approach to be necessary for wolf reintroduction, but they understood its drawbacks. When Fischer, along with Tom France from the National Wildlife Federation, started formulating a strategy for Bitterroot grizzlies, the disadvantages of these tactics were foremost on their minds, which helped them refine and improve their methodology.

Just as national support was increasing, an unexpected twist in the summer of 1987 forced wolf advocates to grapple with the most challenging aspect of managing an apex predator. Early that summer, wolves migrating south from Canada started killing livestock in Browning, Montana, on the east side of Glacier National Park. Throughout the summer, the pack evaded capture, and it seemed that everyone's worst fears were coming true. These incidents fueled the rhetoric of ranchers like Dan Geer, who declared, "If you don't fix this problem right now, I'll kill the wolves myself and hang them on my mailbox, the government be damned." Others were less frenzied, but still wanted to know, "Why should we feed their damned wolves?"— referring to environmentalists, not Canadians. Even though government officials eventually killed or captured all of the Browning wolves, rage against recovery in the West reached an all-time high. Realizing the damaging effect of these livestock losses, Fischer urged Defenders of Wildlife to establish a fund to compensate ranchers for their losses.

By the end of the year, Defenders of Wildlife had raised the funds, reimbursed the ranchers for their summer losses, and the hysteria promptly subsided. This became a critical step and lesson in pushing wolf recovery forward.[20] During this summer-long incident, Fischer, and other wolf advocates, displayed a talent for thinking on their feet and placating opponents, but these efforts were largely reactive, not proactive, and once the hysteria died down, they did little more to accommodate concerns from Old Westerners.

While Fischer described the summer of 1987 as the "Summer from Hell," it did have some positive outcomes. First, the establishment of the compensation fund would be critical to the eventual success of reintroduction. From 1987 through 1994, Defenders of Wildlife would pay out more than sixteen thousand dollars to twenty-one different ranchers for thirty-six cattle and ten sheep killed by wolves. Because the most immediate source of ranchers' aversion to wolf recovery was the potential for livestock depredations, compensating them for live-stock lost to wolves was the most practical way to mitigate this con-cern. Similarly, most of Frank Dunkle's opposition to wolf recovery stemmed from the lack of a good plan to protect ranchers' property rights, but with the compensation plan in place, his resistance waned. National support for wolves continued to grow, and by the end of the summer, the USFWS finally approved the recovery plan, something it had not been willing to do throughout the previous seven years.[21]

Twelve years passed after the Wolf Recovery Team formed before it produced a plan the Fish and Wildlife Service could approve, but this victory was merely a stepping stone to an uncertain future. Once the USFWS had approved the recovery plan, the next step was to secure funds to complete the environmental impact statement (EIS). Required by the National Environmental Policy Act (NEPA) of 1970, the EIS process was designed to ensure that local citizens were aware of how proposed government actions might affect their envi-ronment. It also required the federal government to solicit the views and opinions of the affected public to make certain that any actions were completely transparent and that officials fully considered the ramifications of any plans.[22] The EIS process was costly, however, and in order for it to begin, Congress needed to appropriate the necessary funds. This was much easier said than done.

With the momentum from that summer still fresh, wolf supporters went on the offensive in the fall of 1987 when Congressman Wayne Owens of Utah submitted a bill that required reintroduction to begin within three years. The bill had little effect beyond generating attention for wolf recovery and forcing Congress to address the issue briefly. Two years later, Owens submitted a similar bill that had much wider support and was much more detailed in its stipulations. This bill was destined for the same fate as the first, but before the House of Representatives voted it down, it held a lengthy series of hearings to address the issue.[23]

The following year, in 1990, Jim McClure—a longtime Republican senator from Idaho who had voted for the Endangered Species Act but had declared himself a Sagebrush Rebel and frequently lobbied for the Old West's extractive industries—sponsored a new bill supporting wolf recovery. McClure's support for wolf recovery surprised many people, while a number of environmentalists were downright suspicious. An experienced politician, McClure knew that wolves, like those in Browning a few years earlier, would eventually return from Canada on their own, and when they did, they would have the Endangered Species Act's full protection. This could mean disaster for a number of extractive industries. He had come to realize that it was better to have a say in their recovery than to wait until it was too late. For the first time, McClure's bill recommended *reintroducing* wolves to central Idaho as well as Yellowstone, but because it stipulated that wolves would be delisted upon their reintroduction, most environmentalists refused to support it and the bill died accordingly. Hank Fischer and Defenders of Wildlife had made some attempts to work with McClure as they recognized that he could be their key to success; but after he retired from the Senate in 1991, wolf advocates were forced to proceed without any political support from the delegations of Montana, Idaho, or Wyoming.[24] Even so, McClure's bill kept the channels for discussion about wolf recovery open and encouraged wolf supporters to continue to work toward a solution. In the wake of his bill's failure, McClure helped organize and fund a ten-person Wolf Management Committee to develop a reintroduction and management plan for Yellowstone and central Idaho that would appeal to a variety of interests. The idea was that Congress would more readily fund an EIS if an agreeable plan had already been designed. Composed of agency representatives and members of the public, including representatives from livestock and environmental organizations, the committee had until the middle of May 1991 to make its recommendations. Although committee members fought among themselves over issues such as the ability of private citizens to shoot wolves, the status of wolves as a listed population, and the boundaries governing the experimental classification, the committee eventually agreed upon and recommended a plan. Congress promptly ignored the committee's advice, but that year Congressman Owens sponsored another bill to fund an EIS. This was the fourth year in a row that Congress had considered funding an EIS, so many wolf supporters had lost hope. But, according to Fischer, the Senate Appropriations Committee "had simply run out of reasons for saying no," and it passed Owens's bill. For the first time since the USFWS had approved the recovery plan four years earlier, wolf reintroduction made a significant step forward; and in November 1991, the USFWS began organizing a team to compile the environmental impact statement.[25]

Once the EIS process began, wolf recovery started to progress at a much quicker rate than it had over the previous decade. Opposition from Old Westerners remained high, but NEPA clearly delineated the environmental impact statement process, which allowed it to move forward without added delays. Although the EIS was largely in the hands of the government, wolf advocates continued to press their national media campaigns. Similar to previous battles fought by the environmental movement, the issue was about people, not science, so winning the public relations campaign mattered more than anything else. Over the three-year EIS process, the USFWS held 120 hearings throughout the region, in addition to ones in Washington DC, Seattle, and Salt Lake City, and wolf advocates made sure to take advantage of these highly public events. They turned these hearings into massive rallies based not on facts as much as hoopla and fanfare, and wolf supporters came out in large numbers.[26] The battle was turning in their favor, but just as they gained momentum, a bomb dropped.

The USFWS released the draft EIS in July 1993 and chose as its "preferred alternative" a plan that would reintroduce wolves as an experimental population. But to complicate matters, a hunter killed a wolf two miles outside of Yellowstone's southern boundary that September. The debate over whether wolves still resided in Yellowstone had been simmering for twenty years, but all of a sudden, it was back in full force. Some environmental groups such as the Sierra Club and National Audubon Society, which had become more hardlined in the post-Reagan era and would eventually oppose the ROOTS plan for Bitterroot grizzlies, began to question its legality. Section 10(j) of the 1982 amendment to the ESA allowed the reintroduction of experimental populations only in places where no resident population already existed, because if a resident population did exist, those animals would lose their fully protected status. Using this rationale, the Sierra Club Legal Defense Fund and Audubon Society employed a traditional device of the environmental movement and filed a lawsuit against the USFWS positing that reintroduced wolves should retain their fully endangered status. Wolf supporters—such as Fischer, who had done the majority of the work over the course of the project and had started to see the need for pragmatism—were exasperated. Renee Askins of the Wolf Fund argued, "Laws don't protect wolves, people protect wolves. Greater protection of wolves is not necessarily achieved through more restrictive laws." Fischer added that "the goodwill generated by addressing the concerns of local people would save far more wolves than any number of carefully worded laws." Wyoming's Farm Bureau filed a similar lawsuit, but the courts eventually dismissed both cases on the grounds that a population needed to consist of more than just a few lone animals.[27] Nevertheless, the argument was a significant one that would reappear a few years later as reintroduction of an experimental population of grizzly bears appeared likely in the Bitterroots.

As the courts were deciding these lawsuits, some wolf opponents began to accept the inevitable. Public support at the national level widely favored reintroduction; and the majority of the 160,000 comments the USFWS received following the draft EIS favored reintroduction. Wyoming senator Alan Simpson eventually lamented, "If we're going to have it shoved down our throats, it should be done as an experimental population." By 1994, the Clinton administration had also declared its support for reintroduction, making reintroduction's victory all that more certain. That summer, Secretary of the Interior Bruce Babbitt approved the final EIS. After the courts settled some pending lawsuits in early January 1995, biologists began capturing wolves in Canada, and by the end of the month, Yellowstone and central Idaho had twenty-nine new resident wolves.[28]

While lofty proclamations championing wolves and wolf reintroduction as the embodiment of everything that is good about humans and America had a role in the debate preceding reintroduction, these declarations proliferated in the weeks and months after wolves were released. According to Askins, "The wolf became a vehicle for [grappling with] the profound value shifts occurring in the West." By this estimation, many New Westerners had reason to believe their victory was a signal that their political presence had eclipsed that of the Old West. Representative of this optimism that the West was on the verge of a new political and cultural era was Fischer's declaration that "Yellowstone wolf restoration will stand as a landmark conservation achievement, one that historians likely will use to demarcate the end of an era of predator persecution in the United States." In this same vein, Babbitt added to his earlier sentiments, saying, "It's an extraordinary achievement and it's an important statement about who we are as Americans." Many Americans, New Westerners especially, would have agreed with this notion, but others, Old Westerners in particular, were clearly not as enthusiastic about what reintroduction signaled for the country. The region was changing, but unlike the New West, the Old West did not like what it saw, and it did not agree that wolf reintroduction meant that its values would not ultimately triumph.[29]

While their last-minute lawsuits had failed and wolf reintroduction had finally succeeded, Old Westerners were not ready to hand over control of the region to their new neighbors. Only two weeks after being released in Idaho, one wolf was unlawfully killed near a ranch in Salmon, Idaho. The wolf was found next to a dead calf, but an autopsy confirmed that the wolf had not killed the newborn calf. The experimental designation allowed ranchers to kill wolves caught in the act of attacking livestock, but this incident did not fall within those parameters, clearly making the kill illegal. When USFWS agents went to the ranch where the wolf had been shot to investigate, the seventy-four-year-old rancher greeted the federal agents with hostility, flipping one of the agent's hats off his head and reportedly throwing rocks at them. The incident quickly blew up after the local sheriff criticized the agents for not involving local law enforcement in their investigation. Within a few weeks, Idaho's representatives were calling for a congressional investigation.[30] The episode eventually faded, but it fueled persistent criticisms of a heavy-handed federal government and reinforced complaints made by the Wise Use Movement. Furthermore, it suggested that the debate over wolves was not going to end just because reintroduction had succeeded. And while wolf recovery had no official connection to Bitterroot grizzlies, the two efforts would become symbolically intertwined with grizzlies unable to escape the rhetoric that had emerged during wolf reintroduction.

In addition to two new populations of wolves, the wolf recovery saga had lasting effects on the political and cultural climate of the Northern Rockies, especially as it pertained to natural resource management and predator reintroduction. Tensions between the Old West and the New West had certainly existed prior to the 1990s, but the wolf debate brought them out in the open in the Northern Rockies and forced the two sides into direct conflict. While this conflict had consumed people on both sides, it made some environmentalists, like Fischer and France, begin to question the efficacy of the environmental movement's traditional tactics. While the two steadfast advocates may have heralded the symbolic meaning behind wolf reintroduction, they knew reality was a step or two behind this rhetoric, and they were ready to adapt to that reality while formulating a plan for grizzly bears. Most significantly, Fischer, France, and other likeminded environmentalists were not the only ones to take lessons away from the wolf recovery debate. Idaho's senators Larry Craig and Dirk Kempthorne, Governor Cecil Andrus, and others loyal to the Old West had learned just as much over the course of the process, and their inability to prevent the reintroduction of wolves perhaps made them even more determined to block a new population of grizzly bears.

4

The Advent of the ROOTS Coalition and the Environmental Impact Statement

"No. Hell no!" That was Dan Johnson's initial reaction to the idea of reintroducing grizzly bears to central Idaho. Johnson represented Resource Organization on Timber Supply (ROOTS), which was a nonprofit interest group based in Idaho that consisted of organized labor and timber industry entities dedicated to bringing these groups together to support their collective interests. By the early 1990s, the future of Idaho's timber industry was far from certain. The Forest Service recently had cut its timber allotment throughout the West, and in Idaho, the possibility of new wilderness areas threatened to restrict loggers' access to even more land. In Washington and Oregon, the recent listing of the spotted owl and its critical habitat barred loggers from much of the region's old-growth forests, and in Idaho, the timber industry worried that if the Fish and Wildlife Service added Pacific salmon and bull trout to the endangered species list, similar restrictions would follow. A new population of grizzly bears was the last thing Idaho's timber officials wanted, and ROOTS adamantly resisted any talk of their recovery. At the Interagency Grizzly Bear Committee's (IGBC) annual meeting in December 1993, the Bitterroot Ecosystem Subcommittee planned to present the recovery chapter it had developed over the previous two years, and ROOTS sent Johnson to the meeting with the intention of dissuading the IGBC from pursuing the matter any further.[1]

At the same time, Defenders of Wildlife, a national nonprofit advocacy group dedicated to the protection of plants and animals across North America, sent the director of its Northern Rockies field office, Hank Fischer, to the meeting to express its support for recovery and reintroduction of the bears to the Bitterroots. Fischer had represented Defenders of Wildlife since 1977 and had more experience working for a professional conservation organization than anyone else in the region. He had been involved with a variety of wildlife issues including prohibiting the use of a poison known as Compound 1080, which the government had used to kill unwanted predators. He also contributed to the recovery of the black-footed ferret, and earlier that year he had helped Defenders of Wildlife publish *Building Economic Incentives into the Endangered Species Act*. Most significantly, by 1993 he had been working for more than a decade to recover gray wolves to Yellowstone National Park and central Idaho. And although wolf reintroduction eventually succeeded, his efforts had not yet yielded many substantive results because of the political wrangling that had consumed the issue and slowed it to a crawl. From this experience, Fischer had learned a great deal about achieving environmental reform in the region, and he realized there had to be a better way to achieve endangered species recovery that mitigated the intransigent debates that were plaguing wolves.[2]

Tom France, who worked for the National Wildlife Federation at the regional headquarters in Missoula, was also at the meeting and in a similar position. France had been as intricately involved in wolf reintroduction as anyone, and as an attorney, he not only defended the experimental status for wolves against multiple legal challenges, but he helped shaped the language of the recovery plan. In ensuing years, France has solidified his reputation as one of the most prominent names in conservation in the Northern Rockies, working to restore bison to eastern and central Montana, create more wilderness areas, mandate restoration of open-pit mines, and promote private land conservation. He too had learned a great deal over the previous decade. He wanted to see grizzly bears return to the Bitterroots, but like Fischer, he knew their ability and willingness to work with opponents would be essential to achieving success. Wanting grizzly bear reintroduction to avoid a similar protracted stalemate, they took a chance.[3]

Whereas ranchers who feared livestock predation provided the majority of opposition to wolf reintroduction, Fischer and France knew the timber industry would be the driving force behind any campaign against grizzly bear recovery, and they now understood how essential it was to tackle this problem at the project's outset. Dan Johnson was thinking the same thing. He had not been able to keep the IGBC from approving the recovery chapter, but he still wanted to have a voice in the conversation. And while Johnson and ROOTS did not want to see grizzlies return to Idaho, they had watched the battle between the livestock industry and the environmental community over wolf reintroduction and knew it could be done better. So, after the IGBC approved the recovery chapter over his objections, Johnson found Fischer and France in a nearby bar. He thought they might be willing to work with him and his organization to find a mutually satisfactory solution, and the three men struck up a conversation. After a brief discussion, they agreed their positions on the issue were not entirely incompatible and that additional conversations needed to follow. Although the two sides were still a long way from hammering out the details of a plan, Fischer, France, and Johnson established a foundation, and within a year, they, along with representatives from the Intermountain Forest Industry Association (IFIA), had developed an innovative plan, termed the ROOTS plan, which united certain environmental groups and the timber industry to bring grizzlies back to the Selway-Bitterroot Wilderness.[4]

Although wolf reintroduction ultimately prevailed, the drudgery of the process made Fischer, along with other environmentalists, recognize that the tactics of "traditional" 1960s and 1970s environmentalism were no longer viable in the West. Biologically pure but politically inept plans backed by court decisions and national support networks were no longer practicable. While these tactics had helped attain far-reaching environmental reforms a few decades earlier and had recently undergone a resurgence, they were also largely responsible for the conservative backlash that had hindered environmentalists throughout the late 1970s and 1980s. At the same time, people like Dan Johnson also began to realize the downside of winner-take-all decisions. Idaho's timber industry saw what happened in Washington and Oregon when the spotted owl was listed, and it did not want anything similar to occur in Idaho. With this in mind, Fischer, France, and Johnson shirked their ideological bases and worked with each other to find a way to bring grizzly bears back to the Bitterroots in a way that satisfied both parties. The ROOTS plan was not the first example of this progressive, consensus model, but it was at the forefront of a new wave of natural resource policy and was the first instance in which environmentalists applied collaboration and consensus-based strategies to endangered species recovery. Their innovative plan did not win over every skeptic, but its novelty protected early recovery efforts from the intense debates that had stalled wolf recovery throughout the 1980s and allowed Bitterroot grizzly recovery to proceed with the environmental impact statement (EIS) in a much timelier manner.

For the majority of the 1960s and 1970s, environmentalism had a mandate from the American public that exempted the movement from the political strife that would later consume it. But in Idaho—a relatively progressive state on many issues, especially labor—extractive industries dominated the economy and dictated its political and economic culture. As a result, environmentalism did not gain the widespread acceptance that it did elsewhere. Nevertheless, due to a spirit of compromise, Democratic senator Frank Church, who served Idaho from 1957 to 1981, engineered the formation of vast wilderness and recreation areas throughout the state. He did so not by bullying Idaho's extractive industries, but by forging compromises and placating concerns. While the environmental movement had the power to steamroll any opposition throughout most of the country during this period and was happily doing so, Senator Church brought together

ranchers, loggers, the Forest Service, and environmental organizations such as the Sierra Club. In addition to voting for the Endangered
Species Act, Church helped create 3.87 million acres of wilderness,
two national recreation areas, and 574 miles of wild and scenic rivers in his home state, including the wilderness area that still bears
his name. The battle for each area had its own challenges, but people
respected Church for taking bold action.[5]

Despite the positive reputation he built over his twenty-four years
in Congress, Church and Idaho's progressive culture were no less susceptible to the conservative tide that swept the country at the close
of the 1970s and inspired the Sagebrush Rebellion. In 1980 Church
lost his fourth bid for reelection. In the years following his depar-
ture from Congress, the spirit of cooperation and compromise for
complex natural resource issues that he had championed throughout
his career deteriorated. As the Sagebrush Rebellion gained momentum, attempts to enact environmental legislation in Idaho and the
West "produced a bitter and intense political polarization" that would
persist into the 1990s and shape the discussion over grizzly bears.[6]
Church's time in office produced lasting protections for Idaho's vast
public land system, but when efforts toward Bitterroot grizzly recovery began in earnest, any lessons learned from Church's accomplishments seemed like distant memories.

Even before Bitterroot recovery started moving forward, the grizzly bear recovery program as a whole had attracted some negative
attention, especially from the timber industry. Each national forest
was required to regularly revise its forest plan, which denoted how
that forest would be managed, and during this process, each forest
considered what lands would be logged and what lands might be
better suited for other uses, like wildlife habitat, recreation, or wilderness. Because grizzly bears were threatened, the agency had to
conduct section 7 consultations to take into account how any actions
would affect them, and these consultations regularly forced the Forest Service to place greater restrictions on logging. Naturally, the
timber industry, which was still reeling from the restrictions brought

by the USFWS listing the spotted owl, were not happy that another threatened species was reducing their allotments. As a result, forest plan revisions turned into high-profile political battles in which grizzly bears became scapegoats for Old Westerners who believed they were being driven out of the forests.[7] In Montana, the Flathead National Forest became a battleground between loggers and grizzlies; and in Idaho, these debates played out over the revision of the Panhandle and Kootenai National Forest plans. Grizzly bears may not have been as controversial as wolves, but their role in the "Timber Wars" did not help their popularity among many in the West, including politicians.

Old West loyalists were not the only ones displeased with the state of grizzly recovery in the lower forty-eight states. Environmentalists were just as critical of how the Fish and Wildlife Service had been handling recovery efforts because they believed the service had not been doing nearly enough to protect the bear or its habitat. "They've blown it. Hundreds of thousands of acres have been clear-cut or explored for oil and gas," claimed Mitch Friedman of the Greater Ecosystem Alliance. When the USFWS revised the Grizzly Bear Recovery Plan in the early 1990s, some environmentalists complained that its recovery zones and target populations were too small. In 1993 a group of environmental groups that included the Greater Yellowstone Coalition, Wilderness Society, and the Sierra Club threatened to sue the USFWS to reclassify the Selkirk population of bears as endangered instead of threatened.[8] Protestations from the environmental community not only forced the government to tighten codified protections for bears, but increased tension between conservative Old Westerners and the federal government. Grizzly bear recovery had been progressing and by that measure it was starting to prove successful, but politically, the program was rife with turmoil. These matters did not directly affect recovery in the Bitterroot ecosystem, but because these debates played out in close proximity, Bitterroot stakeholders were familiar with the common issues and concerns surrounding grizzly bear management, and when Bitterroot recovery became a major

topic of debate, both sides were well versed in the standing arguments for and against.

Nevertheless, early efforts to return grizzly bears to the Bitterroots
were moving along at a steady pace with few hindrances. The U.S. Fish
and Wildlife Service first approved the Grizzly Bear Recovery Plan in
1982, and after a handful of habitat studies conducted throughout the
1980s confirmed that the Bitterroots remained well suited for grizzlies, the Interagency Grizzly Bear Committee authorized the compilation of a Bitterroot chapter in late 1991 and added it to the GBRP.
Early the next year, the IGBC appointed Wayne Wakkinen of Idaho's
Department of Fish and Game as the team leader for the Bitterroot
Grizzly Bear Working Group—the committee responsible for writing
the recovery chapter. The working group included biologists from the
USFWS, the Forest Service, Idaho Fish and Game, and Montana Fish,
Wildlife, and Parks, but Wakkinen received the nod as team leader
so that Idaho, where Bitterroot grizzly recovery would be most controversial, could maximize its input. Wakkinen was already in charge
of recovery in the Selkirk ecosystem, so he had valuable experience
behind him, which made him a good fit for the job.[9]

The working group first met in February and in June decided to form
the Citizens Involvement Group, which was comprised of a diverse
segment of the natural resource community of Idaho and Montana
and would provide a connection between the working group and the
public. Only seven people attended its first meeting in August, but by
the end of the year the group's membership had grown to thirty, and
Wakkinen had to limit its size to make its meetings manageable. In
fact, Wakkinen was forced to cancel one meeting because after groups
from both sides rallied their constituencies, the meeting threatened
to devolve into a raucous shouting match. After Wakkinen restricted
the group's size, it was able to operate effectively over the remainder
of the process. Neither Hank Fischer nor Dan Johnson attended the
first meeting, but after Wakkinen extended invitations to them in
August, they became eager and consistent participants even though
they found little to agree upon at first.[10]

As with the environmental impact statement process that followed, the working group drafted the recovery chapter, held public comment meetings, and had a general public comment period before presenting the chapter to the IGBC. The working group produced its first draft by August 1992, and in September it held public meetings in Hamilton and Missoula, Montana; Lewiston, Grangeville, and Orofino, Idaho; and Salt Lake City to give the public an opportunity to discuss the issue. The atmosphere at most of the meetings was tense as various people—many of whom were still fighting to stop wolf reintroduction—expressed fears that grizzly recovery would mean threats to personal safety and economic livelihoods. There was also a great deal of misinformation that had circulated around some of Idaho's small towns that led people to believe the USFWS wanted to reintroduce two hundred bears to the region. In spite of these charges, Chris Servheen, the grizzly bear recovery coordinator for the USFWS, who oversaw the entire grizzly bear recovery program, tried to mitigate everyone's concerns. He insisted that reintroduction would take place on a much smaller scale, no land-use restrictions would accompany the bears' recovery, and the threat of an attack would be minimal. Furthermore, he maintained, "When the real truth comes out, it's not near as bad as people have heard." Many people refused to accept these assurances. "I've talked to guys in Libby or Columbia Falls, Montana, guys in my industry who say you don't want grizzlies over there," said Earl Britt, an Idaho timber worker. The crowds seemed on edge, but the overall mood of the meetings was relatively tame, and the working group considered them a success as they were well attended and helped educate many people on what to expect.[11]

While the working group attempted to incorporate suggestions voiced during the public comment period into the final draft of the chapter, Idaho was attempting to maximize its role in the process and exclude the federal government entirely. In its first session of 1993, Idaho's House of Representatives introduced a bill that would create a nine-person Grizzly Bear Management Oversight Committee to advise Idaho's Department of Fish and Game on grizzly bear policy. In the spirit of the Wise Use Movement's calls for a limited federal government, the bill intended for the committee's decisions to trump any made at the federal level, but it did not provide *any* funding for management. Rather, it forced the state's grizzly bear managers to rely solely on the $140,000 it received annually from the federal government. In reaction to the bill, Jerry Conley, the director of Idaho's Fish and Game, said that it would effectively cut Idaho out of federal decision making and have little effect because the federal government still had jurisdiction. Hank Fischer charged that it would put power in the hands of bureaucrats instead of wildlife professionals and added, "You can't say we want control and then don't do anything." Others asserted that Idaho did not need the committee because it already had plenty of input and warned, "If state involvement is too restricted, Idaho will lose control of bear recovery to the federal government." In spite of these warnings, the House passed the bill sixty-four to two, and the Senate followed suit.¹²

The Bitterroot Grizzly Bear Working Group continued to refine its recovery chapter, but Idaho's attempt to interfere with the process exposed the project to new criticisms. The chapter was not a plan to reintroduce bears, but local citizens could easily see what was coming next, and Idaho senator Larry Craig cautioned that the chapter "charts a course of disaster." Furthermore, fears concerning safety and economic stability that arose during the previous summer's meetings escalated. Because many Idahoans were still fighting to keep wolves out of their state, it did not take much for the mood surrounding grizzly bears to go from concerned to frantic. Referring to the restrictions that would supposedly accompany any reintroduced grizzlies, logger and Clearwater County Republican Central Committee chairman Pat Richardson roared, "We think the radical environmental community couldn't find a spotted owl so they decided to import one and they settled on the grizzly bear." Although the listing of the spotted owl had not directly affected Idaho's extractive land users, it was a lesson they did not take lightly, and combined with an existing disdain for the federal government, it made many Idahoans suspicious of any similar actions. In that same vein, Richardson chided that the issue was not about bears but an attempt to create de facto wilderness. Idaho governor Cecil Andrus added, "Just because they can't find any bears there, why does that mean we have to have more?" He further questioned whether grizzly bears were even endangered. Another commentator suggested, mocking the supposed spiritual and ecological benefits that grizzlies would bring, "The only thing the grizzly bear adds to the forest is fear." Building off that idea, a Lewiston man unapologetically stated, "I don't like the idea of humans being part of the food chain." Finally, some even suggested that if the federal government did reintroduce bears, Idahoans would manage them using the SSS method—shoot, shovel, and shut up.[13]

While many Idahoans were debating *whether* to recover grizzly bears in the Bitterroots, others continued to debate *how* to recover them. Some opposed reintroduction but were not necessarily ready to block recovery. The Back Country Horsemen of Idaho supported natural recovery and opposed any restrictions on land use; they would relent to reintroduction only if the recovery goal was set at a low number. The Clearwater Board of Commissioners wanted the recovery area—the region in which the USFWS would focus its efforts and have the authority to manage bears as a priority—to include only the Selway-Bitterroot Wilderness. ROOTS agreed with this revision as well. While the group first and foremost did not want recovery to proceed, it insisted upon an experimental population restricted to the Selway-Bitterroot Wilderness if it did take place. ROOTS had been involved with the Citizens Involvement Group since 1992, and although it found the group's meetings fruitless, it continued to take part in the discussions in order to maximize its input and ensure that no additional restrictions on logging resulted.[14]

How to define the exact boundaries of the recovery area proved to be one of the most persistent problems that plagued the working group. Unlike wolves, grizzly bears required immense areas of relatively intact habitat to survive. Not only did bears roam over great distances yearly and throughout their lifetimes, but the numerous foods they relied upon—such as whitebark pine nuts, army cutworm moths, newborn elk, and deer calves—existed in vastly different habitat conditions, and grizzly bears needed all of them. So as controversial as delineating the recovery areas for wolves had been, grizzly bears needed much more land that was much less impacted by modern infrastructure, which increased the stakes of designating their recovery area.

The draft version of the recovery chapter extended the recovery area's boundary south to the Salmon River, but the forest supervisor of the Salmon National Forest wanted the boundary moved north because he believed habitat in the southern region was poorly suited for grizzly bears, and historically, grizzlies did not inhabit the region. The draft plan had also extended the boundary eastward, a few miles into Montana, on the edge of the Bitterroot Valley. But the Bitterroot National Forest's supervisor wanted the boundary to be moved to the Idaho-Montana border so that it would include only the Selway-Bitterroot Wilderness—in order to decrease the chances of bears coming down into the heavily populated valley. Finally, the Lolo National Forest supervisor wanted the recovery chapter to exclude certain areas that received heavy recreational traffic because he believed they would become areas of frequent human-bear conflict. The working group considered these suggestions, but it also contemplated extending the recovery area farther south beyond the Salmon River to include the entire Frank Church–River of No Return Wilderness Area in order to situate this new population of bears as close as possible to the Greater Yellowstone ecosystem (GYE).[15]

Although some wanted to restrict the size of the recovery area, many environmentalists wanted to expand it as much as possible. As the USFWS finalized the Bitterroot recovery chapter over the summer and fall of 1993, the only significant populations of grizzly bears that existed in the lower forty-eight states were in the Northern Continental Divide ecosystem (NCDE) and the GYE. The NCDE supported more than three hundred bears, and roughly two hundred and thirty bears roamed the GYE, while fewer than seventy-five bears existed in the Northern Cascades, Cabinet-Yaak, and Selkirk ecosystems combined. Additionally, since biologists began tracking bears in 1975, no grizzly had migrated between ecosystems, even though fewer than fifty miles separated four of the island regions. The lack of a grizzly migration history between regions led supporters of reintroduction—in response to those who advocated natural recovery through migration— to stress its biological improbability. Furthermore, this separation had started to worry biologists who were concerned that genetic isolation of these distinct populations would be detrimental to their long-term viability. This concern made recovery in the Bitterroots essential in the eyes of some advocates because the mountain chain was part of the largest roadless area in the lower forty-eight states and was situated between the NCDE and GYE. If a healthy population of grizzlies could be restored to the Bitterroots, then hope existed that bears would migrate and solve the problem of genetic isolation on their own. For this reason, the committee deemed it necessary to extend the recovery area boundary as far south as possible. However, others, such as Chris Servheen, dismissed the idea that a population in the Bitterroots would lead to migration between populations, as it had never happened previously. He argued for reintroduction because, logically, it was the most timely and efficient way to satisfy the Endangered Species Act's requirement to recover bears in the Bitterroot ecosystem.[16]

Because the public comment process had drawn increasing negative attention to grizzly bear recovery, the working group, which had been renamed the Bitterroot Ecosystem Subcommittee, wanted to remove as many of the chapter's controversial aspects as possible and leave those issues to be decided during the environmental impact statement process. However, environmental organizations such as Wild Forever—an umbrella organization comprised of the Greater Yellowstone Coalition, Sierra Club, Wilderness Society, and the Audubon Society and dedicated itself to protecting grizzly bears—charged that the subcommittee made this decision only because it was "getting squeamish about facing the rough political climate in central

Idaho." More specifically, they castigated the USFWS for favoring the experimental, nonessential status too strongly in the recovery chapter, wanting to delay the decision on the boundaries of the recovery area, and wavering on its commitment to require reintroduction. Adam Ruben, of Wild Forever, reminded the subcommittee that the purpose of the chapter was to establish the biological needs of bears, not to placate political opposition. He warned that if the group delayed decisions concerning the recovery area's boundaries and reintroduction until the EIS, political pragmatism would eventually trump biological necessity and the bears would suffer as a result.[17] Ironically, Wild Forever changed some of its positions over the ensuing years, but these concerns would prove prophetic.

In spite of these admonitions, the final version of the chapter the subcommittee presented to the IGBC in December 1993 did not strictly delineate the recovery area's boundaries. It did, however, state that recovery would require reintroduction, and it redacted language to which some environmentalists had objected that suggested reintroducing only four to six bears. Finding no major objections to the chapter, the IGBC approved and added it to the GBRP along with a recovery chapter that had been developed independently for the North Cascades.[18] The project to return grizzly bears to the Bitterroots had taken two years to get a recovery chapter approved, whereas wolf recovery had needed twelve years to reach that milestone. The established bureaucratic channels in the form of the IGBC were partly responsible for this, but the inclusive spirit fostered by the Citizens Involvement Group also helped streamline this process. And even though resistance to recovery had a hand in influencing the finalized chapter, the project was off to a running start.

Americans have long been enthralled by pragmatic politics, and while the practice was undergoing a nationwide revival by the end of the twentieth century, the concept was relatively new to the field of natural resource management. The passage of the Wilderness Act in 1964, NEPA and the Clean Air Act in 1970, the Clean Water Act in

1972, and the Endangered Species Act in 1973 all represented changing generational values, not the marriage of divergent groups. As
discussed in chapter 1, this legislation was effective because it was
legally binding, not because the entire country inherently agreed
with the principles behind it. When environmentalists believed that
an agency violated these laws, they pursued the matter through the
courts, as the battles to save the snail darter, spotted owl, and gray
wolf demonstrated.

In the 1990s, much of the environmental community was turning
more hard-lined, but the conservative backlash to environmentalism,
championed by the Sagebrush Rebellion and its later iterations, made
some environmentalists realize the limits of their political influence
and forced them to moderate their approach. Although some environmental organizations were championing the biologically pure solutions of the 1960s and 1970s once again, others began to recognize
the need for pragmatism. This led them to adopt a consensus strategy,
which they viewed as a way to avoid the large expense of lawsuits;
make better, quicker decisions; and return some degree of civility to
the process. At the same time, extractive industries, especially the timber industry, which faced multiple threats to its long-term prosperity,
began to find ways to accommodate environmentalists' concerns. As
the spotted owl incident proved, taking a no-compromise approach
could have devastating consequences, and the timber industry could
no longer risk the unpredictability of winner-take-all decisions that
had the potential to lock them out of the forests completely.[19]

However, by the time the IGBC approved the Bitterroot recovery
chapter, only a handful of collaborative coalitions existed across in the
country. The most prominent of these was the Quincy Library Group
(QLG). The QLG formed in Plumas County, California, in 1992, when
the owner of a local sawmill approached a local forest watch group
about developing a mutually beneficial forest plan for the surrounding national forests. The group gained immense local support, but
when it presented its plan for approval, the Forest Service rejected it.
Eventually, the group pushed its plan through Congress, past objec-

tions from environmentalists who argued that it superseded legally established regulatory policies; and in 1998, President Clinton signed it into law. However, this controversy did not make national news until 1997 and had yet to receive much national press.[20]

Other cooperative coalitions were also taking shape throughout the West at the beginning of the 1990s, but none of these groups tackled issues on the same scale as returning a fearsome predator to the country's largest roadless area. In 1993 the Henry's Fork Watershed Council formed in Island Park, Idaho, to bring together recreationists, who used the Henry's Fork for boating and fishing, and farmers who needed the river's water to irrigate crops. The two groups had been incapable of cooperating; but through the council, they were able to enact conservation measures that allowed farming to continue in a more sustainable manner. In the Swan Valley, north of Missoula, an ad hoc committee had come together to protect the timber industry and the area's natural resources. The ad hoc committee's decisions were nonbinding, but because of its diverse participation, it quickly gained respect and influence over a range of decision makers in the valley. A final example of consensus building had its start in 1993 when Secretary of the Interior Bruce Babbitt released his Rangeland Reform '94 initiative, which intended to alter long-standing grazing policies across the West. This policy infuriated extractive land users who accused Babbitt of launching a "War on the West." In reaction to this impending debacle, a group of ranchers and environmentalists proposed creating local citizen advisory councils to review each region's grazing plans. The idea won acceptance, but it was not instituted until 1995–so when Fischer, France, and Johnson, along with representatives of the Intermountain Forest Industry Association, began meeting in the spring of 1994, they were on the cutting edge of natural resource management policy.[21]

By the fall of 1993, when the fate of wolf reintroduction was still uncertain, Fischer and France started writing letters to the House and Senate Appropriations Committees requesting fifty thousand dollars to commence the EIS process for Bitterroot grizzlies. From their experiences with wolf reintroduction, Fischer and France knew that if the project experienced delays, it would become not only more controversial but exorbitantly more expensive. Even though they wanted to accelerate the process, they knew that the success of grizzly reintroduction, like that of wolf recovery, would hinge on human values first and science second. If people wanted grizzlies and were willing to tolerate them, they would succeed. If they did not, grizzlies would be as nonexistent in the Bitterroots as they had been for the previous fifty years. Fischer and France understood they had to craft a plan that local people would support. Grizzly bears reproduced slowly, so even a few human-caused mortalities could threaten the population's viability. Therefore, it was especially important that any grizzlies introduced to the Bitterroots avoid the SSS style of "management."[22]

As Fischer said, referring to the process of wolf reintroduction, "I came away feeling pretty frustrated. It took so long; it was so expensive; it was so polarized. While we may have won the battle, I'm not sure we won the war." Wolf reintroduction was a couple of years off, but Fischer had already learned from the experience. With this in mind, Defenders of Wildlife launched a grassroots campaign to manufacture public support, sending out mailings to local environmentalists requesting their financial and physical assistance, conducting surveys and making public presentations in towns throughout Idaho.[23] Additionally, Fischer and France wasted little time following up with Dan Johnson after their first conversation in Denver because they knew that a relationship with Johnson and ROOTS could be the link that would make grizzly bear recovery happen.

Before ROOTS began working with Defenders of Wildlife and the National Wildlife Federation, the group had emerged as a leader among organizations that did not want to see grizzly bears return to central Idaho. As of late 1993, ROOTS opposed recovery in the Bitterroots—questioning both its biological necessity and whether the region's habitat could support a population of bears. However, taking a lesson from Senator Jim McClure, the group knew that if grizzlies made it to the Bitterroots on their own, they would have the Endangered Species Act's full protection. They still opposed reintroduction, but with this in mind, they altered their official position, which forecasted their willingness to find compromise. ROOTS's revised statement read, "If the IGBC insists on recovering grizzly bears in the S-B [Selway-Bitterroot] ecosystem, then we propose it do so through introduction of a nonessential, experimental population within the existing boundary of the S-B Wilderness."[24]

During meetings held by the Idaho Grizzly Bear Management Oversight Committee in the summer of 1993, a number of people expressed support for ROOTS's new stance, including Idaho's Democratic congressional representative Larry LaRocco, other timber industry leaders, wildlife biologists, and surprisingly, the oversight committee itself.[25] As a result, ROOTS had positioned itself as a major player in Bitterroot grizzly recovery discussions even before it began working with Defenders of Wildlife and the National Wildlife Federation.

Even though ROOTS had plenty of support from conservative interests, it was still open to working with Defenders of Wildlife and the NWF, and by May 1994 it had made a commitment to working toward a solution that would bring grizzlies back to the Bitterroots. A month earlier, the three organizations had met in Missoula, which was the first since meeting Fischer, France, and Johnson spoke in Denver. More than a dozen ROOTS members attended the meeting, which was brief; but that June, they sat down for their first serious discussion in Orofino, Idaho. In addition to Fischer and France, Mike Roy from the NWF, Phil Church, a union rep from the Potlatch timber mill in Lewistown, and Bill Mulligan, a mill owner and member of ROOTS, were all in attendance.[26]

At that meeting the group went over a booklet that the NWF had produced, which outlined grizzly bear history and biology and provided information on how grizzly bear recovery would affect local citizens. As grizzly recovery had already proved to be an emotional issue, not always grounded in accurate information, Fischer had insisted prior to the meeting, "It strikes me that the most fundamental place to start is from a common base of facts." The groups decided to tackle one of the trickier issues next—the boundaries of the recovery area. Johnson and ROOTS had already made it clear that they wanted recovery to be restricted to the wilderness area, but the NWF and Defenders of Wildlife had also made requests during the development of the Bitterroot chapter to expand the recovery area to include public lands outside of the Selway-Bitterroot Wilderness. This discussion grew tense at times as no template existed for what these groups were attempting, but nothing was off the table, and the discussion was open and honest. All sides left the meeting in good spirits, ready to schedule their next gathering.[27]

These first two meetings were well-attended, especially by members of ROOTS and other timber interests; but learning from the experience of wolf reintroduction, Fischer and France wanted to maximize participation. They knew that in order to reach true consensus on grizzlies, they had to be as inclusive as possible because anyone who had a stake in grizzly bear recovery would eventually make their voice heard, and it was better to have that input sooner than later. In this spirit, they invited to their next meeting members of the Clearwater Forest Watch Coalition, Idaho Conservation League, Wild Forever, and Sierra Club—groups that often took hard-lined environmental stances and had already demonstrated they were unwilling to make any substantial concessions on grizzly bear recovery.[28]

Reliant on their battle-proven tactics, honed since the advent of the environmental movement forty years earlier, none of these groups was willing to deviate from the most biologically pure approach to recovery. As a result, none of them responded to these invitations, and Defenders of Wildlife and the NWF ended up being the only major environmental groups involved with developing the ROOTS plan. The coalition did add to its ranks, however, the Intermountain Forest Industry Association, an Idaho-based organization dedicated to representing the timber industry's interests. The group was generally wary of all grizzly recovery efforts, and the association's president, Jim Riley, was hesitant to work with this coalition because of possible land-use restrictions, but when the IFIA's wildlife representative, Seth Diamond, pushed the matter, Riley relented. Diamond had worked for the Forest Service on a number of endangered species issues, and when he came to the IFIA in 1994, he had the biological know-how to lead the way. Soon after joining the coalition, Diamond started promoting ROOTS's position to other Idaho forest users, and he would be one of its most valuable and effective members.[29] With this diverse alliance intact, the four groups launched a grassroots campaign throughout Idaho and Montana to build public support while they continued to meet, discussing and refining their plan to return the iconic bear to the Bitterroots.

The mood of the coalition's meetings was often anxious as the groups debated the necessity of grizzly reintroduction and the areas into which the bears would be reintroduced, but they continued to make progress. By the end of September, ROOTS released a revised position statement. Whereas they had prefaced earlier statements with "We do not want grizzly bears recovered in the Selway-Bitterroot Ecosystem," their revised statement simply stated, "We support recovery within the existing boundary of the Selway-Bitterroot Wilderness." However, ROOTS insisted on the nonessential, experimental designation, and neither Fischer nor France was able to convince ROOTS to expand the recovery area. Even so, the group had agreed upon an extended experimental boundary in which managers would tolerate bears, but downgrade the degree to which they would be managed as a priority. ROOTS's new position statement also advocated habitat studies in the Frank Church–River of No Return Wilderness to determine the suitability of its habitat for bears.[30]

Finally, the group had started to outline what would be the most innovative, as well as most controversial, aspect of its plan—citizen management. Responding to the most persistent criticism from advocates of the Property and County Rights Movements as well as opponents of wolf reintroduction—the lack of local control that accompanies top-down federal environmental action—the ROOTS

coalition wanted to find a way to let local people direct decision making.[31] Since the mid-1960s, environmentalists had relied on the federal government to enforce its reforms, but in the spirit of cooperation,
Defenders of Wildlife and the NWF were willing to move away from
this model. The coalition was a few months away from hammering out the details of this aspect of the plan, but the progress it had
made over the course of the year was remarkable and groundbreaking nonetheless.

At the beginning of 1994, after the IGBC added the Bitterroot chapter to the recovery plan, Fischer, France, and others in their organizations had initiated a campaign to request funds from Congress to
start the EIS process. Once the ROOTS coalition solidified, however,
they pursued the matter with added vigor. One of the major obstacles to wolf reintroduction had been the lack of political support, but
due to this unlikely partnership that moderated the traditional "environment vs. economy" debate, politicians were much more willing
to support funding for a grizzly bear EIS. The immediate reasons
Old Westerners gave for opposing environmental reforms typically
revolved around concern for the effects that added regulations would
have on extractive industries. They complained that environmental
reforms were top-down measures that infringed upon local auton-
omy; but the unique nature of the ROOTS coalition mitigated these
concerns and alleviated cause for alarm.

At the prompting of the ROOTS coalition, Larry LaRocco (of Idaho)
and Pat Williams (of Montana) wrote a joint letter to the chair of
the House Appropriations asking for EIS funding. This was relatively unprecedented and demonstrated a distinct break from what
politicians were willing to do for wolf recovery. Pat Williams had
side-stepped the wolf debate as much as possible because it was too
controversial, and in 1988, he had recommended against the augmentation of the grizzly population in the Cabinet-Yaak ecosystem
because it was unpopular among locals. LaRocco first took office in
1991, so he did not have as much of an opportunity to take part in the
debates over wolf recovery, and once in office, he astutely avoided the

issue. However, because a variety of interests expressed support for grizzly bear recovery, both were willing to write letters and campaign for it. Joe McDade (of Pennsylvania) and Julian Dixon (of California), both on the Appropriations Committee, also wrote the chairman asking for funds. Furthermore, Montana's Fish, Wildlife, and Parks Department, Idaho's Department of Fish and Game, and even the Idaho Grizzly Bear Management Oversight Committee, all expressed approval for moving forward with the EIS process. Republicans made huge gains in the midterm elections that November, which empowered a much more conservative Congress, but despite this turn in political tides, ROOTS and the IFIA never wavered in their commitment. As Bill Mulligan of Three Rivers Timber said, "We're sticking with this process because it's the right thing to do."[32] They helped build this coalition in an attempt to move away from the obstinacy that typically plagued environmental debates, and despite doubts from the environmental community concerning the sincerity of their commitment, they did not waver.

Meanwhile, by 1994, the Democratic Clinton administration had finally taken a firm stance in support of wolf reintroduction, which momentarily dampened opposition to Bitterroot grizzly recovery. Additionally, the Fish and Wildlife Service and the IGBC had been making their own preparations to move forward with the EIS process under the assumption that it would pursue reintroduction rather than natural recovery. Chris Servheen estimated that the entire NEPA process would take roughly eighteen months, cost just under $250,000, and be finished by the end of 1996. As a result of this wide-ranging support, on January 9, 1995, a few days before wolves returned to Yellowstone and central Idaho, and just as the Wyoming Farm Bureau and the Mountain States Legal Foundation were filing last-minute petitions to prevent their release, Congress appropriated funding for the EIS.[33] Despite its inauspicious timing, the announcement garnered little attention, and grizzly advocates were able to transition smoothly into the EIS process. One might suppose that because one controversial wildlife issue was forefront in the news, Congress would be hesitant to take any bold steps concerning another. But the timing of this announcement could not have been better testament to the ability of the ROOTS coalition to reach the level of consensus that had eluded wolves, even on the eve of their release.

Only four years had passed from the time the IGBC decided to pursue recovery in the Bitterroots until the EIS process began. It had taken wolf recovery sixteen years to get that far, and cost $6 million as opposed to $500,000. By all measures, Bitterroot grizzly recovery was moving at lightning speed, and thanks to the ability of the growing ROOTS coalition to break away from traditional environmental tactics and strategies and embrace a consensus approach, many had reason to believe that grizzlies would be back in the Bitterroots in just a few short years. Defenders of Wildlife president Rodger Schlickeisen seemed justified in his estimation that Bitterroot grizzly recovery would "become one of the great endangered species success stories of the next decade." This accelerated timeline was due in part to the lessons some environmentalists had learned from their experiences during wolf recovery. Wolf recovery had been a long, hard fight, and even though it was struggling across the finish line by January 1995, Fischer and France began to realize that this old model of effecting environmental reform no longer worked. They wanted grizzly bear recovery to avoid a similar two-decade-long battle, and to do this, they revived the tradition of uniting divergent interests to compromise on natural resource issues originally championed by Frank Church.[34]

As a result of this collaborative effort, Fischer and France and their timber industry cohorts found themselves leading a burgeoning movement that championed compromise and sought to end the partisan strife that had impeded environmental reform measures since the late 1970s. But if they thought their innovative partnership and their experience during wolf recovery was enough to escape or alleviate resistance from the Old West (or the New West), they were in for a big surprise. Many environmentalists remained content with the 1960s and 1970s approach to environmental activism that mandated biologically pure decisions supported by the federal government and backed by the courts. And even though the ROOTS plan had the support of the timber industry, the Old West was more powerful and entrenched in the region than any of its singular components, and its resolve and determination to maintain its influence would not be broken. The ROOTS coalition had good reason to be optimistic, but the region's political stalemate would not be defused so easily.

5

Environmental Resistance

By 1995 the influence of the Missoula-based environmental organization Alliance for the Wild Rockies (AWR) and its cofounder, Mike
Bader, was on the rise. Formed just a few years earlier, the AWR
was committed to protecting the Northern Rockies including Wyoming, Idaho, Montana, Washington, Oregon, and into Canada. They
championed the bioregional approach to environmental reform that
ignored political boundaries such as state lines and roads that nor-
mally demarcate national forests, national parks, or other public
lands and conservation districts. Instead, they relied on more natural boundaries like rivers, mountain ranges, and habitat types to
determine proper boundaries, and they treated the entire Northern
Rocky region as a single, connected ecosystem. Over the previous few
years, they had made their mark by campaigning for the Northern
Rockies Ecosystem Protection Act (NREPA), which would have designated twenty-four million acres of new wilderness across the five
states, with nine and a half million acres and seven million acres in
Idaho and Montana, respectively. The act would have ended federally
subsidized timber harvests, mining claims, and grazing allotments
and restored migration corridors for wildlife.[1] The act would have
ushered in a wholesale shift in how land was managed in the West
that reflected a much different vision for the region—one in which
environmentalism and the New West reigned supreme.

With this idealistic bill as their chief goal, the AWR exemplified the type of local, hard-line environmental organization that had regained power in the post-Reagan era and forced larger groups like the Sierra Club to shift to the left. Their attempts to turn the bill into law so far had been unsuccessful, but they had earned a reputation that made them a force within the environmental community. Not surprisingly, certain groups believed their vision was too radical and unrealistic. Organizations like Defenders of Wildlife and the National Wildlife Federation, which may have agreed with the principles of the NREPA, did not believe hard-line policies like this were realistic or could be implemented. Because of this, an ideological divide began to pit New West environmentalists against one another, with one side choosing pragmatism over idealism

Riding a wave of popularity engendered by growing support for the NREPA, the AWR took a defiant stand against the ROOTS plan in the debate over Bitterroot grizzlies, believing the proposal made far too many concessions on behalf of the bears and did not do enough to protect habitat. While the Sierra Club, Audubon Society, and Greater Yellowstone Coalition may have had greater national name recognition, the AWR had the resources and know-how to spearhead this campaign and formulate an alternative that would have recovered the Bitterroots' grizzly population in a much different manner. Today, however, the AWR is a far smaller organization that does not have the same level of popular appeal or name recognition. The NREPA is still floating around Congress, coming to the floor most recently in 2013, but the momentum behind it has withered.[2] Mike Bader's role in the environmental community has also waned since the mid-1990s, and when I first called him in 2012 for an interview, the idea of Bitterroot grizzly recovery sounded like a long-forgotten memory.

Before helping form the Alliance for the Wild Rockies, Bader worked as a park ranger in Yellowstone, where his duties included bear research and management. Grizzly conservation had always been an important issue for him, but today his strongest connection to the grizzly appears to come from Mike Bader Bearjam—the blues/ funk band he leads. Even so, when I sat down to talk with him, the issue seemed as fresh in his mind as it was nearly twenty years ago. Prior to getting involved with Bitterroot grizzlies, Bader had tracked the debates over wolf recovery closely, and he firmly believed that politics had trumped biology in the end. He understood the realities of the social and political climate, but because of grizzlies' slow reproductive rates, he wanted to ensure that biology would lead the discussion over Bitterroot grizzlies, and he was determined to fight the experimental designation. At the same time, he believed that while groups like Defenders and the National Wildlife Federation wanted to capitalize on the momentum created by wolf recovery's success, the region was not ready for a new population of grizzlies. According to Bader, "We can honestly say wolves don't attack people and eat them, but we can't say that about grizzlies." He added, referring to the restrictions and precautions that would have accompanied a new population of grizzlies, "People are just set in their ways and they don't want to change. They think change is probably bad, especially if the government is suggesting it."[3]

With this in mind, Bader confirmed that he is happy the project failed, because he still distrusts the idea of citizen management and thinks the experimental designation would have proved disastrous. And he trusts that when the issue eventually resurfaces, it will have the chance to succeed or fail on its own merits and not be unfairly influenced by the wolf issue.[4] Despite this firmly stated belief that the region was not ready for a new population of grizzly bears, when Bader and the AWR realized that Bitterroot recovery was moving forward, they actively participated in the debate, pushing for what they considered the most biologically pure solution. As the environmental impact statement process mandated by the National Environmental Policy Act proceeded, rhetoric from Bader and the AWR escalated, but Bader insists that they were mostly there to play devil's advocate. If this was the case, their campaign was an unparalleled success, because their participation fundamentally influenced the debate. Their opposition helped delay the release of the environmental impact statement and undermined the project's chances for success.

Even though Bader believed that Defenders of Wildlife and the NWF were unwisely using their victory in wolf reintroduction as a platform from which to pursue grizzly recovery, Fischer and France joined the ROOTS coalition in order to avoid the political ineptitude that had delayed wolf recovery. Perhaps even more so than Bader, they were aware of the divided nature of the Northern Rockies' political climate, and they believed grizzly recovery could avoid those same controversies. However, while Hank Fischer, Tom France, Dan Johnson, and Seth Diamond were able to step away from their ideological bases in order to accomplish this, most people involved in the debate, including many environmentalists, were not. The ROOTS coalition wanted to move the EIS process forward as quickly as possible, but two and a half years passed—after Congress appropriated the funds—before the USFWS even released the draft EIS. During this time, the ROOTS coalition continued to respond to criticisms and refine their plan, but they spent valuable time and resources bickering over its details with environmentalists who shared a similar vision. Because these hard-line environmentalists exemplified the New Western mindset, their participation in the debate would come to mean that, ironically, intransigence on the part of the New West would bear a considerable amount of responsibility for the project's ultimate collapse.

Immediately after receiving funding from Congress, in January 1995, the Interagency Grizzly Bear Committee and the USFWS began organizing the recovery team and taking the needed steps to begin the environmental impact statement. First, the Bitterroot Ecosystem Subcommittee appointed John Weaver as the recovery team leader. Weaver had worked on the wolf recovery team and proved to be a strong member, willing to stand up to extractive industry interests for the benefit of wolves. Also, he had been the grizzly bear habitat coordinator for the Forest Service during the 1980s, so his political and scientific experience made him a sound choice for the position. Similar to the group that developed the Bitterroot chapter four years earlier, the recovery team included members from the USFWS, Forest Service, Idaho Fish and Game, Montana Fish, Wildlife, and Parks, and Nez Perce Tribe, whose reservation bordered the west side of the recovery area. By spring the team had laid out a schedule that anticipated the release of the final EIS by summer 1996, with a final ruling being made by that fall. At its February meeting, some participants noted that the political climate had already grown tenser in the preceding months. But because the team knew the environmentally friendly Clinton administration would be in office for only the next four years, it intended to adhere to the hastened schedule to ensure the project would not be subjected to the whims of a less wildlife-friendly administration.[5]

While the Bitterroot recovery team was ramping up its efforts, Chris Servheen made the decision to halt recovery in the North Cascades. Washington still had a few remnant bears, so recovery would very possibly have meant augmentation as opposed to reintroduction. Proponents of recovery in the Cascades were quick to point out that this would be politically easier to accomplish than reintroduction. They also advocated for the Cascades on the basis that the population was genetically unique and to lose it would be tragic. But funding was limited, and the idea of being able to connect the Greater Yellowstone ecosystem to the Northern Continental Divide ecosystem was alluring. Servheen had only enough funding to support one project, and because of the ROOTS coalition's early success building bipartisan support, Bitterroot recovery seemed like the best bet.[6]

Over the next few months, members of the recovery team, especially Weaver and Servheen, attended meetings throughout Idaho and Montana, including with Idaho's legislature, individual federal and state representatives, and others who wanted a voice in grizzly bear recovery. Perhaps most notably, they met on a number of occasions with Hank Fischer, Tom France, Dan Johnson, Seth Diamond, and other members of the ROOTS coalition to help refine the plan. The recovery team wanted to see Bitterroot grizzly recovery succeed, and it was aware of the political fight that reintroduction would produce, so ROOTS's groundbreaking compromise quickly gained favor. The team knew that political support for the ROOTS plan was largely responsible for Congress funding the environmental impact statement so quickly, and it wanted support from these groups to continue. This political reality put the ROOTS plan in a strategically advantageous position to become the preferred alternative in the draft statement, and even early in the process it emerged as the leading proposal.[7]

By spring the ROOTS group had hammered out many of the plan's details, beginning with the recovery zone and the experimental recovery area. The groups expanded the recovery area to include the Selway-Bitterroot and Frank Church–River of No Return Wilderness Areas, encompassing 3.7 million acres; but reintroduction would occur only in the Selway-Bitterroot Wilderness, and the recovery goal would be based only on the holding capacity of that area. However, the revised plan also included an experimental population area, in which bears would be tolerated, but not managed, as a priority. It would be 15.3 million acres, slightly smaller than West Virginia, and extend in all directions from the recovery zone. Reintroduction would take place over a five-year period and include a minimum of twenty-five bears. Most significantly, the plan created an eight- to ten-person steering committee to be appointed by the governors of Idaho and Montana and comprised of diverse interests. The committee would be in charge of policy-level decisions concerning the bears in the experimental area, while the respective state wildlife agencies would have the primary responsibility over day-to-day management. Finally, the plan explicitly stated that grizzly bear recovery would not hinder economic development and that public education would be required for people to coexist peacefully with grizzly bears.[8] The plan was taking shape, but the group knew it was not finalized, and over the next five years, it would continue to refine the plan and try to make it as palatable as possible to its diverse spectrum of opponents.

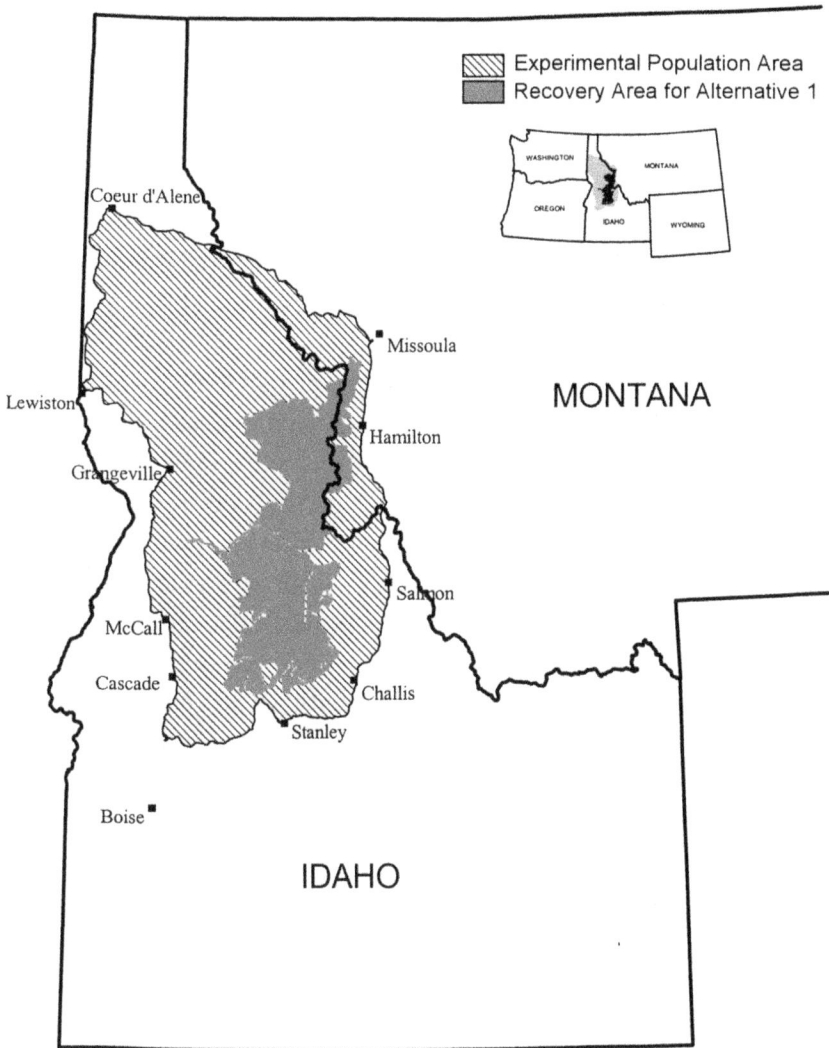

Map 3. Map of the recovery area under the ROOTS plan. The recovery area under the
ROOTS plan included the Selway-Bitterroot and Frank Church–River of No Return
Wilderness Areas, but the experimental population area stretched across a much
larger land mass. U.S. Fish and Wildlife Service.

Despite the progress Fischer and France had made with their timber industry allies, other environmental groups had shown little interest in taking part in the coalition, believing, as one environmentalist did, "When you crawl into bed with the enemy, you become the enemy."[9] This statement exemplified the fanatical loyalty that New West environmentalists felt for their ideology and the polarization that had occurred during the early 1990s. Even so, Fischer and France still believed they could gain the endorsement of their environmental cohorts.

In March they sent a letter to thirty organizations and members of the environmental community detailing their progress within the ROOTS coalition and assuring doubters of the timber industry's sincerity and commitment to recovering bears. They stressed the fact that the EIS had moved forward with much less fanfare than the wolf EIS because of the timber industry's support and argued that the timber industry had much at stake in making sure grizzly recovery happened successfully. In closing, Fischer and France reminded their colleagues that collaboration was a give-and-take process and that they had made huge strides toward reintroduction over the previous year.[10]

Some recipients responded favorably to the appeal and understood the importance of what Fischer and France were doing. Dennis Baird of the Sierra Club thought the plan could have been stronger with its protection for grizzlies, but given the political climate following the reintroduction of wolves, he considered it "courageous and visionary." Others had specific suggestions, such as basing the recovery goal on the holding capacity of both wilderness areas, instead of just the Selway-Bitterroot Wilderness, and including the northern region of the experimental area as part of the recovery zone. However, these critics recognized that the positive reception following the announcement to prepare the EIS had resulted from the trust built during the collaborative process, and they urged Fischer and France to continue working in that spirit.[11]

Despite these encouragements, a coalition representing the

National Audubon Society, Idaho Conservation League, Sierra Club, Greater Yellowstone Coalition, and the Wilderness Society sent a response that expressed their "grave concern" that the coalition's principles set "dangerous precedents." Some of these groups had been involved with the lawsuit to keep reintroduced wolves' status as fully endangered, and they believed the ROOTS plan was not "sound conservation" and would not lead to the bears' recovery. Even though Fischer and France stated that the experimental status and citizen participation were cornerstones of the plan, these environmental groups were hostile to those provisions. They believed that the non-essential, experimental designation incorrectly implied that a Bitterroot population was not vital to the species' long-term survival. Similarly, they did not think the plan did enough to create and protect linkage zones between recovery areas and encourage migration. They also attacked the principle of the citizen management committee because it was nonscientific. According to them, partisan politics would steer the committee, and states would illegally manage federal lands. According to Matthew Reid of the Great Bear Foundation, allowing Idahoans to manage grizzlies "borders on insanity." Finally, they suggested that ROOTS's commitment to making no economic impact was unrealistic and too restrictive at that early stage in the process. These groups offered their support for the idea of collaboration and consensus, and Mike Clark of the Greater Yellowstone Coalition stressed the need for continued dialogue, but their inflexible devotion to science prohibited them from allowing any substantive concession to be made on the bears' behalf and prevented them from taking part in any meaningful collaboration.[12]

In contrast to their theoretical reception of collaborative and consensus-based process, these groups remained outwardly hostile to the realities of compromise, and they were not alone. Although many people across the West were singing the praises of collaborative conservation by the mid-1990s, many others openly doubted its benefits and remained happily wedded to the tactics of 1960s and 1970s environmentalism. Some doubted the legality of measures that took authority away from federal agencies and put them in the hands of private citizens. Others believed that collaboration was merely a tool of industry designed to cripple standing regulations and erode strict standards set by legislation. In 1992 the Forest Service, Bureau of Land Management, timber industry, and some environmentalists had formed the Applegate Partnership in Oregon to create a forest plan for the surrounding national forests. Environmentalists initially praised their efforts, but some eventually soured toward the group because in its efforts to establish authority, the group allegedly violated the National Environmental Policy Act by not allowing enough chance for the public to voice its opinion. The controversy that eventually embroiled the Quincy Library Group (see chapter 4) and brought it into the national spotlight was similar in nature. Many of its detractors charged that the process moved so slowly because any agreements and subsequent actions required trust, which could not be built quickly among groups that had opposing interests. Still others believed that politics drove collaborative decision making much more than did science. Accompanying this criticism was the idea that by politicizing an issue, groups with more political clout gained disproportionate influence, which usually hurt nonprofit environmental organizations and the natural resource at issue and benefitted the large industries that long held clout in the political process.[13]

Despite these criticisms, the ROOTS coalition had built momentum since the previous year and in addition to holding meetings with the USFWS and the recovery team, they had met with citizens and local officials in towns like Hamilton, Orofino, and Grangeville. Through the end of 1995 the ROOTS proposal was the only Bitterroot recovery plan, but the hard-line environmental groups were persistent. Instead of just attacking the ROOTS plan, these groups understood that they needed to make their voices heard in a more positive, constructive manner.

While larger groups like the Sierra Club and Audubon Society had led this faction at first, by the end of the year, Mike Bader and the Alliance for the Wild Rockies emerged as the most active voice. Although the AWR was a much smaller organization, its work with the Northern Rockies Ecosystem Protection Act afforded it the necessary resources and influence to take charge. Much like its allies had done, the AWR initially entered the debate by attacking the ROOTS plan. During the summer of 1995, it circulated a news release that accused the timber industry of forming the ROOTS coalition as part of a "campaign to sabotage the reintroduction" and scolded Defenders of Wildlife and the NWF for being "all too eager to accept reintroduction at any cost." But by the end of the year, the AWR realized it needed to produce a recovery plan of its own to influence the debate in a serious and constructive manner. Over a three-week period, beginning in December 1995, Bader and Timothy Bechtold, of the Ecology Center, worked tirelessly, and by January, they released the conservation biology alternative (CBA) as a response to the ROOTS proposal.[14]

The CBA exemplified the tenets of bioregionalism and hard-line environmental reform that had come back into fashion over the previous few years. First, the plan called for the primary means of restoration to be natural migration. Bears would have to make it back on their own, but when they did, the emigrant bears would retain the full protection of the Endangered Species Act, unlike a reintroduced population, which would be downgraded to nonessential, experimental. To make natural migration possible, the CBA explicitly banned future road building in the recovery zone and required massive habitat restoration to create and maintain linkage corridors between the Bitterroots and recovery areas to the north. The ROOTS proposal, in contrast, did not address road building and did not prevent it outright. The CBA allowed translocation to expedite the bears' return if necessary, but because it envisioned reconstructed migration corridors, it mandated natural migration as the principal mode of returning grizzly bears to the Bitterroots.

Because it maintained the bears' fully threatened status, the CBA required section 7 consultations for any policy changes that could affect the ecosystem to ensure that grizzly bears would not be negatively impacted. Under the nonessential, experimental designation,

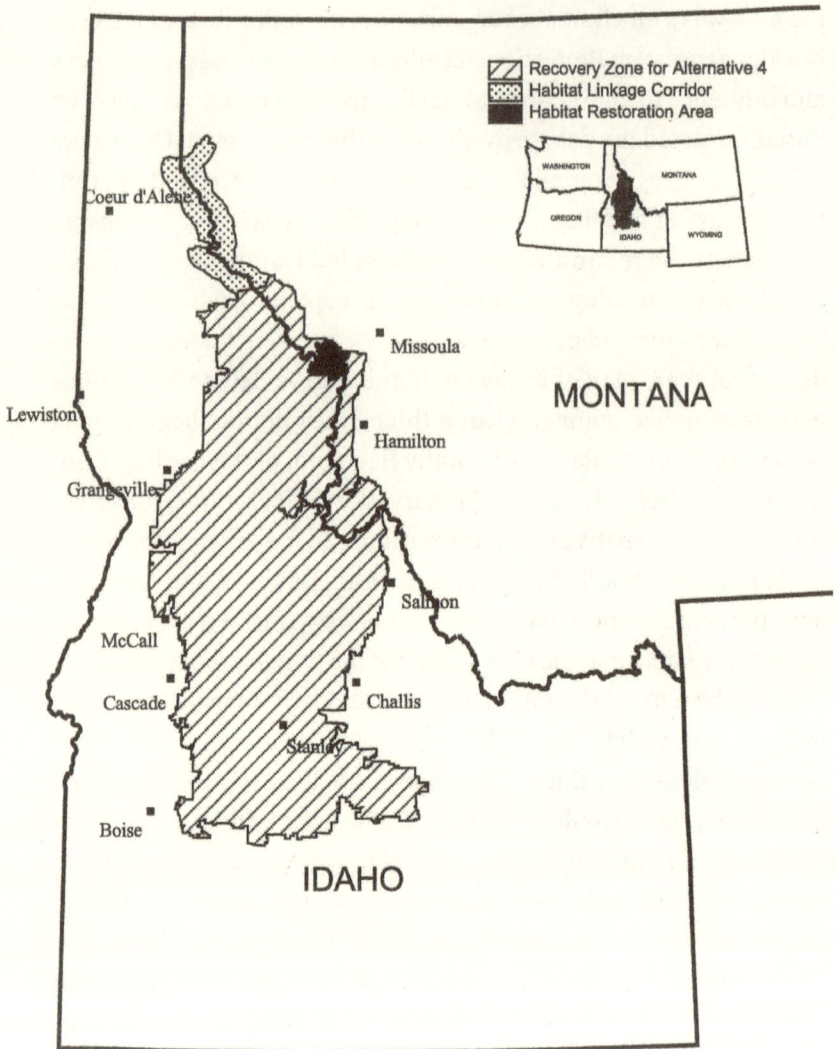

Map 4. Map of the recovery zone under the conservation biology alternative. The recovery zone under the conservation biology alternative, where grizzly bear management would be a priority, was significantly larger than that of the citizen management alternative and included a habitat linkage corridor to connect the Bitterroot and Cabinet-Yaak ecosystems. U.S. Fish and Wildlife Service.

section 7 consultations were not required, so this provided another safety net to protect the bears' habitat. Next, whereas the ROOTS proposal contained a recovery zone and an experimental recovery area, the CBA featured only a much-expanded recovery zone. The recovery zone in the ROOTS plan was 3.7 million acres, while the CBA's recovery zone covered 13.8 million acres and was slightly smaller than the area covered by the experimental population area in the ROOTS plan. Additionally, instead of a citizen management committee with the authority to make binding decisions, the CBA created a ten-member scientific committee to serve purely in an advisory role. The Fish and Wildlife Service would retain full decision-making authority. The ROOTS plan allowed the government to issue permits to citizens to kill bears that were attacking livestock, but the CBA authorized citizens to kill bears only in self-defense. Finally, the CBA claimed that by converting old roads to their natural state and preventing subsidized logging on federal land, it would create twelve hundred additional jobs over a ten-year period and save taxpayers sixty-nine million dollars.[15]

Before Bader and the AWR compiled the conservation biology alternative, resistance from environmentalists to the ROOTS plan was haphazard and defined in negative, not positive, terms. The release of the CBA, however, codified their position and organized their opposition. Even after this occurred, the ROOTS coalition focused most of its attention on gaining the support of conservative adversaries, but the release of the CBA undoubtedly escalated the debate between the two sides. ROOTS's proponents responded to the release of the CBA with an editorial by Michael Roy of the National Wildlife Federation. He couched most of his objections to the CBA in terms of how unrealistic a proposal it was, and he accused Bader of settling "into a comfortable position from which he can snipe while hiding behind the guise of science." Bader had questioned the legality of the nonessential, experimental designation, but Roy confirmed that the Bitterroots met all three requirements to be eligible for that status—historic occupancy, isolation from other populations, and no resident population. Bader had also criticized reintroduction because he said no surplus bears were available to be translocated. Roy reminded Bader that while Canada—from which the ROOTS plan proposed to take its bears for reintroduction—did not necessarily have a surplus, any bears taken from that population would be counted toward Canada's annual hunting quota. As a result, it effectively saved bears from hunters' bullets. On that same point, Roy pointed out that biologists had already agreed that reintroduction was the only realistic way to engender recovery as no migration between recovery areas had previously occurred or was likely to occur.[16]

Next, in response to the AWR's criticism that the ROOTS plan would not require section 7 consultations and would not do enough to protect habitat, Roy fired back that the other species listed in the Selway-Bitterroot region, such as Pacific salmon and Canada lynx, would still require section 7 consultations, and what was good for those species was good for grizzlies as well. Also, because wilderness areas made up the 3.7-million-acre recovery zone, the region was already off-limits to logging and road building. Furthermore, as Tom France added, the citizen management committee would have input into any changes made to forest management plans, so it would be able to lobby on the bears' behalf. ROOTS proponents also compared the road densities for the national forests in the Bitterroot ecosystem to those in the Northern Continental Divide ecosystem. In the Clearwater and Nez Perce National Forests, the road density was only one-half mile of road per square mile whereas the NCDE, where grizzlies were doing well, had one mile of road for every square mile. Additionally, they emphasized that in both the Clearwater and Nez Perce Forests, far fewer acres had been logged than their forest plans' allotted. France believed that the CBA's economic analysis was thus flawed because it was based on inflated numbers. Finally, ROOTS advocates pointed out that the Forest Service had built only thirteen miles of new roads in those areas since 1994 while it had removed 247 miles.[17] For them, the ROOTS plan did not need to ramp up protection for habitat explicitly, which would have been politically risky, because the region's habitat was already being administered in a way that was conducive to grizzly management.

Seeming to justify conservative charges that added land restrictions would result, other ROOTS members countered criticisms that the plan did nothing to create linkage corridors by confiding to their environmental colleagues that a prior designation of these corridors would lessen the chance of reintroduction. According to their way of thinking, they would be in a stronger position to bargain for them after bears were on the ground. Bader later dismissed this assurance saying, "That's what drives people nuts. . . . You promised us you wouldn't do that but now you're here for a second bite at the apple." For Bader, one of his plan's greatest attributes was that it was up front and contained all the restrictions and guarantees he wanted, and no one could accuse him of trying to hide his intentions—a charge made frequently against environmentalists.[18]

Finally, Roy scolded Bader, saying that both science and politics were important when introducing grizzlies because if people did not accept the plan, it would fail. "The question isn't habitat quality. It's simply if we want them to survive," quipped recovery team member Steve Nadeau. Roy acknowledged that the ROOTS plan had a lower biological chance of recovering grizzlies, but he argued that political realities had to be considered, and he called Bader's plan simply "impossible to implement." In this same vein, Tom France charged that the proposal erroneously assumed that support existed for radically revamping how national forests would be managed.[19]

In retrospect, Bader admitted that his plan had little chance of being implemented, but at the time, he fought for it with a zeal that suggested he had higher hopes for it. Either way, he wanted to make sure that it was part of the discussion and that science was not forgotten in the rush to find compromise. If it came down to it, he preferred no introduction to a compromised one, believing that the region was not ready for a new population of grizzlies and the slowly reproducing bears would suffer as a result. In this light, some believed that Fischer and France had sold out to industry. Mark Solomon of Wild

Forever accused them of being "pseudo conservationists."[20] More substantively, Bader chided, "If it's worth bringing an endangered species into central Idaho, it's worth doing it right." By January 1996, wolves had been in central Idaho and Yellowstone for a year, and already, they were surpassing the expectations that many had for their recovery at such an early stage. Many advocates of the ROOTS plan pointed to wolves in addition to other examples of nonessential, experimental success stories—such as black-footed ferrets, California condors, and whooping cranes—as proof that the experimental designation would adequately protect bears as well. However, supporters of the CBA cautioned ROOTS members that the breeding habits of wolves and bears were vastly different, and while wolves might be well-suited for such a status, bears were not. "Experimental status tends to get abused .... People cheat, the government cheats ... shooting bears when they don't really need to," suggested Jim Olsen of the pro-bear group, Friends of the Bitterroot.[21] And because a grizzly population was more fragile, it could not sustain even minimal losses as wolves could.

While grizzly bears indeed presented unique challenges to wildlife managers, ROOTS supporters had reason to be heartened by other examples of experimental, nonessential populations across the country. By the late 1970s, scientists believed the black-footed ferret to be extinct, but when a rancher spotted one outside of Meeteetse, Wyoming, the U.S. Fish and Wildlife Service rushed to recover what they knew must have been a rapidly dwindling population. By capturing ferrets and breeding them in captivity, the USFWS began reintroducing experimental populations in pockets from Canada to Mexico, as well as many states in between. By the late 1990s, recovery was already proving fruitful, and since then, the program has continued to prosper with an estimated one thousand wild ferrets by 2013. The California condor has also thrived under experimental, nonessential designation. The population had declined over the twentieth century until it reached a low point in the 1980s, at possibly no more than twenty-two worldwide. In 1987 researches took the remaining wild condors into captivity, and by 1996, after intensive breeding efforts, the USFWS established experimental, nonessential populations in northern Arizona. While this program would have been too new for anyone involved with Bitterroot grizzlies to evaluate, California condors have done relatively well, with roughly 240 wild birds between California, Arizona, and Mexico and 435 worldwide by 2013.[22]

Whooping crane recovery has followed a slightly different path, and while its recovery appeared successful by the close of the century, certain crane populations have not been able to maintain their numbers. Whooping crane populations declined precipitously over the nineteenth and early twentieth centuries due to habitat loss and overhunting, and the federal government added the bird to the endangered species list in 1967. In the early 1990s, the USFWS started reintroducing an experimental, nonessential population in Florida, and initially the population exploded, increasing from eight in 1993 to eighty-seven by 2001. Unfortunately, it has since declined and by 2011 was back down to twenty-five. The Rocky Mountain population has fared even worse. In the late 1990s, the USFWS removed its critical habitat and deemed it experimental, nonessential; and by 2002, the agency officially acknowledged that no population inhabited the region. Other populations have thrived, and nearly six hundred whooping cranes exist worldwide, but the plight of the Rocky Mountain population undoubtedly offers a cautionary tale.[23] No one involved with Bitterroot grizzly recovery could have known for certain how the great bear, with its unique set of challenges, would have fared under the experimental designation, but at the time, that question was the focus of the debate between ROOTS supporters and proponents of the CBA.

Hard-line environmentalists were intensely critical of the experimental designation, but they were even more critical of citizen management, which they believed should be strictly advisory. They maintained that the citizen management committee would not provide a scientific voice and politics would trump science, especially considering that the governors of Idaho and Montana would be responsible for recommending the majority of the committee. Some environmentalists simply did not like the idea of these states having control over the bears' management. They added that by allowing logging to continue, the plan would undermine the Sweet Home decision, a recent Supreme Court case that confirmed the illegality of harming a listed species' habitat. Responding to the idea that the forest plans would provide sufficient protection because of the other endangered species present, Brian Peck of Wild Forever charged that the forest plans in the experimental area were not currently compatible with grizzly management. Among these groups, there were also persistent whisperings that the experimental designation was illegal because a few grizzly bears still roamed the Bitterroots.[24]

Although the ROOTS coalition tried to assuage the concerns of their environmental colleagues by convincing them of their plan's merits, they also continued to refine the proposal to make it more palatable to them. By the end of 1995, the revised ROOTS plan mandated that the citizen management committee would base its decisions on the best scientific data available and encouraged it to solicit advice from experts in the field. It also ensured that the secretary of the interior would be required to review and approve a work plan the citizen management committee would submit every two years. It expanded the committee to include fifteen members—seven recommended by Idaho's governor, five recommended by Montana's governor, one appointed by the secretary of agriculture, one from the secretary of the interior, and one recommended by the Nez Perce Tribe. These members would all serve six-year terms, and the plan required the committee's meetings to be open to the public. Environmentalists feared that extractive industry would dominate the committee, so if the committee's actions were not leading to the bears' recovery, the revised ROOTS plan included a clause that allowed the secretary of the interior to force a judicial review and assume full control if necessary.[25]

Even with these amended provisions, hard-line environmentalists still refused to support the proposal, and they continued their attacks on the ROOTS plan. Members of the coalition were exasperated. They had to balance the needs of their timber allies with the habitat needs of the bears for the plan to succeed, and they were getting no help from their colleagues. In a letter to Tom France and Michael Roy, Hank Fischer pleaded, "It doesn't benefit anyone except the opposition when conservationists attack each other." But despite Fischer's pleas, this bickering continued. Seth Diamond sardonically noted, "If grizzly bear biologists can't agree, of course citizens won't agree." Even more to the point, Mike Roy commented that there were more grizzly bear experts than actual grizzly bears.[26]

The ROOTS coalition wanted to keep the channels of communication open with the hope that their environmental colleagues would eventually support the plan, but the window for productive conversation was quickly closing. Attacks from the left got even nastier and surpassed a simple disagreement over the means of achieving a similar goal. Louisa Willcox of the Greater Yellowstone Coalition and Wild Forever tried to undermine Fischer's credibility with Defenders of Wildlife by leveraging her connections with the organization's board of directors. When ABC News ran a story on the reintroduction in 1997, Fischer got word that Mike Bader told a reporter that Fischer was a "renegade" who did not have support from his organization. The vitriol reached an apex at the Alliance for the Wild Rockies' annual meeting in 1996. It had become a tradition to announce the "Golden Stump Award," given to the person who had done the most harm to the environment over the previous year. That year, the "award" went to Hank Fischer and Tom France.[27] While the two sides certainly had different visions of how grizzly recovery should occur, this action demonstrated how rancorous and irrational the debate between people who generally had the same priorities had become. While hard-line environmentalists may have believed that the ROOTS plan made some unwise compromises, to claim that Fischer and France had been the environment's biggest enemies was hyperbole beyond reason that undermined their credibility.

The recovery team had originally planned to release the final EIS by June 1996, but by the middle of that year, eighteen months after Congress had allocated funding, the draft was far from complete, and the Fish and Wildlife Service was considering not completing the EIS at all. Following the midterm elections in 1994, Congress had swung to the right significantly with Republicans making substantial gains in both the House and the Senate. The USFWS was prepared to have its budget cut by twelve million dollars. Wolf reintroduction had received a major boost once the Clinton administration pledged support, but the administration had yet to throw its weight behind Bitterroot grizzlies. It had been willing to stand up to this conservative backlash on behalf of wolves because of the years of work that had gone into it and because of how close reintroduction was to becoming a reality. Compared to wolves, grizzly reintroduction was in relative infancy. Additionally, the threat of litigation made by hard-line environmental organizations that promoted the CBA and were hostile to the collaborative approach made supporting any grizzly recovery plan more politically risky for the administration. President Clinton had developed a good reputation within the environmental community, and the fact that so many environmental groups opposed the ROOTS plan made his administration hesitant to support it. With all of this uncertainty swirling about, the USFWS contemplated cutting the budget for the EIS from $250,000 to $150,000. As he did for wolf reintroduction, Bruce Babbitt eventually pledged his support and the funding came through. Still, damage had been done. Babbitt's support was more for the EIS process than for the ROOTS plan itself. The administration did not want to undermine the NEPA process, but it also feared losing support from powerful environmental organizations like the Sierra Club.[28] As a result, the veil of sweeping consensus had lifted, and with some in the New West just as critical of the plan as their Old West counterparts, the project was open to partisan debate more than ever.

In many ways, Fischer and Bader were cut from the same cloth. Both originated from outside of the West—Fischer from Ohio and

Bader from Iowa. Each had moved to Montana to pursue a career in the natural resource management field and to take advantage of the region's recreational opportunities. They both attended the University of Montana. Bader received a bachelor's of science in resource conservation, and Fischer earned a master's degree in environmental studies, for which he had studied wildlife biology and journalism. After school, Fischer joined Defenders of Wildlife, while Bader chose to work as a ranger in Yellowstone. By the end of the 1980s, Fischer was still with Defenders of Wildlife, and Bader left the Park Service to enter the environmental advocacy world with Alliance for the Wild Rockies. Fischer had more experience in the field, but the two men found themselves traveling a similar path. They were both prototypical constituents of the New West who exemplified the values of this burgeoning demographic. But from here, their paths diverged. Whereas Fischer began to understand that compromise was needed to accomplish environmental reforms in the West, Bader refused to bow to political pressure. These differing visions did not change the fact that both men fell solidly within the parameters of the New West, but they came to represent competing visions of this growing constituency.[29] Which model would the New West follow—dogmatically sticking to its principles and trying to overthrow the current order, or adapting to the complex nature of the system and working toward reform that was realistic in the short term? New Westerners had not previously had to make this choice, but as grizzly reintroduction proceeded and collaboration and consensus grew in popularity over the closing years of the century, these would become important questions that New Westerners would have to answer.

Regardless of the direction the New West has taken in this regard, thirteen years after the project collapsed, Bader still believes that his organization's role in the debate over grizzlies will ultimately prove beneficial for the bears. Because of their different breeding habits, wolves' success under the experimental designation has had little effect upon his conviction that such a designation would have been bad for bears. Bader maintains the comparison of citizen management to letting "the fox manage the chickens," and he is optimistic that a biologically pure solution to Bitterroot grizzly recovery could be realized in the near future. Ultimately, he is proud of the disruptive role that he and his organization played during the debate.

By the time the project fell apart at the hands of the Bush administration, blame most easily could be placed on the shoulders of the Old West's unyielding devotion to its principles. However, unwillingness on the part of certain New Westerners to find compromise within their own ranks led to significant delays over the first couple of years of the NEPA process, which had as much to do with the project's failure as the refusal of the Old West to break away from its ideology. The ROOTS coalition continually tried to respond to and ease the fears of their environmental colleagues by revising the plan to ensure that the reintroduced bears and their habitat would receive the necessary security. Still, these revisions and assurances fell short of the demands being made, and because an influential segment of the environmental movement had turned more radical over the previous few years, hard-line environmentalists were opposed to making any biological sacrifices for the sake of political realities.

Fig. 1. Mission Mountains from the top of Cha-paa-qn. From Cha-paa-qn, the Mission Mountains stand less than thirty miles to the north and host a robust population of grizzly bears. Photo by the author.

Fig. 2. Bitterroot Mountains from the top of Cha-paa-qn. From Ch-paa-qn, the northern reaches of the Bitterroot Mountains are less than twenty miles to the south, yet contain no viable population of grizzly bears. Photo by the author.

Fig. 3. Seth Diamond, Hank Fischer, and Dan Johnson enjoy a moment relaxing. Seth Diamond of the Intermountain Forest Industry Association, Hank Fischer from Defenders of Wildlife, and ROOTS representative Dan Johnson worked together to bring grizzly bears back to Idaho and Montana under an innovative set of guidelines. Getty Images.

Fig. 4. A grizzly bear wanders through the Greater Yellowstone ecosystem. For some people, grizzlies inspire curiosity and invoke emotional attachments to the natural world. For others, they are a source of fear that have no place living beside humans. These divergent opinions were the source of the Bitterroot controversy. Photo by Barrett Hedges.

Fig. 5. (*opposite*) Hamilton meeting room during public comment hearing. In Hamilton, the idea of reintroducing grizzly bears to the Selway-Bitterroot Wilderness, which was visible from town, inspired passionate reactions and helped fill many meeting rooms over the course of the project. Photo by Tom Bauer for the *Missoulian*.

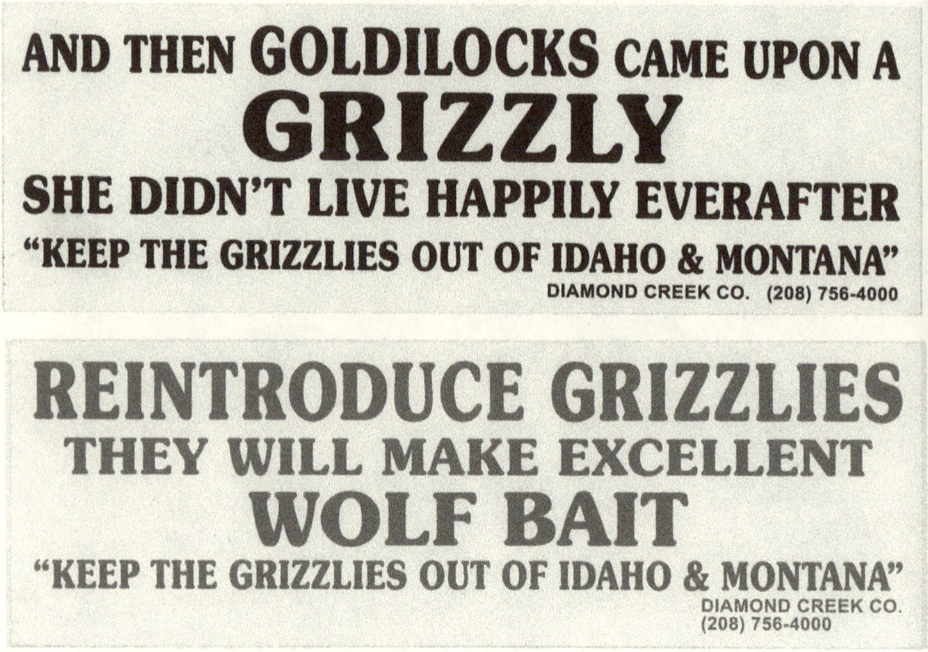

AND THEN **GOLDILOCKS** CAME UPON A
GRIZZLY
SHE DIDN'T LIVE HAPPILY EVERAFTER
"KEEP THE GRIZZLIES OUT OF IDAHO & MONTANA"
DIAMOND CREEK CO. (208) 756-4000

REINTRODUCE GRIZZLIES
THEY WILL MAKE EXCELLENT
WOLF BAIT
"KEEP THE GRIZZLIES OUT OF IDAHO & MONTANA"
DIAMOND CREEK CO.
(208) 756-4000

Fig. 6. Anti-grizzly bumper stickers. On both sides, the debate often relied more on emotional rhetoric than it did on facts. Photo by the author.

Fig. 7. Sculpture of a grizzly bear in Salmon, Idaho. The sculpture depicts two species that were the center of controversy in the 1990s. Despite its prominent presence in the town, residents passed a resolution in 1999 asserting that grizzlies never inhabited the region. Photo by Dan Flores.

Fig. 8. Sculpture of a grizzly bear in Missoula, Montana. Representations of grizzly bears appear around Missoula and across the state. Acceptance of the bear as a symbolic figure greatly overshadows the degree to which Montanans celebrate its real-life counterpart. Photo by Zack Porter.

6

Ethical Controversies and the Draft Environmental Impact Statement

The meeting room was crowded and restless when Bitterroot Valley resident Dennis Palmer rose from his seat and declared, "We don't want the doggone bears." This bold declaration, while representing the sentiments of many attending the public meeting in small, conservative Hamilton, Montana, was more measured than others. Long-time resident Robert Norton confidently stated, "Women and children are going to be killed and maimed." During a similar meeting held in Hamilton two years later, an opponent of grizzly reintroduction read aloud the pathology report of a woman who had been mauled by a grizzly bear in Glacier National Park and displayed a picture of the mangled body for everyone to see. Another positively asserted that people "would rather reintroduce rattlesnakes and water moccasins than grizzly bears." Histrionics reached an apex when one local resident lifted his young daughter above his head in the middle of the meeting room. Everyone's eyes turned toward the young girl as her father announced to the room that she would be bear bait if the federal government reintroduced grizzlies. In *High Country News*, a reporter summed up the frenetic atmosphere of a meeting in rural Salmon, Idaho, which was similar to the other six meetings held across Montana and Idaho in October 1997, by wryly observing, "Big, stout, fully grown men displayed the kind of hostility and fear bordering on panic that, when voiced by women, is usually dismissed as hysteria."[1]

Passions ran just as high on the other side as people who supported recovery felt no need to pull their punches in the public meetings, which gave citizens of the region a chance to voice their opinions on the U.S. Fish and Wildlife Service's draft environmental impact statement for Bitterroot grizzly recovery. In a meeting held in Boise, a supporter of the great bear taunted his opponents, "If you want a sterilized, artificial environment, go to Disneyland." He continued, "I'd expect it [fear of bears] from industry, but it really disgusts me to hear hunters, backpackers or whatever say they're scared." Although the atmosphere at most of these meetings was raucous and frenzied and even bordered on chaotic, the room was silent in Missoula, Montana, when the eight-year-old granddaughter of renowned grizzly bear biologist John Craighead stood at the microphone in front of two hundred attentive onlookers and made an impassioned, heartfelt plea for the bears. Back in Hamilton, one speaker tried to rally support for reintroduction by emphasizing a sense of "moral responsibility," while another pleaded for the bears' restoration by couching the issue as a matter of national, not local, concern. If these meetings affirmed one thing, it would be Chris Servheen's claim that "nobody has no opinion about grizzly bears."[2]

By the 1990s, the country's environmental ethics were split into two camps—conservationism and environmentalism—and the ROOTS group, which came to refer to itself half-jokingly as the "radical center," faced attacks from both sides. To make matters more challenging for the innovative coalition, the country's politics at large were becoming increasingly divided, and fewer and fewer people were willing to make any compromises on grizzly bears. People who adhered to conservation's wise-use ethos were largely Old Westerners, and their criticisms included concerns for public safety, fears of land-use restrictions that could accompany the new population of bears, high costs, and a belief that an additional population of grizzlies was unnecessary. From the left, New Westerners, fanatically loyal to the principles of the environmental movement, condemned the architects of the ROOTS plan for not doing enough to protect the bears' habitat or the bears themselves. Both positions enjoyed significant public support, but the Fish and Wildlife Service made the ROOTS plan its preferred proposal nonetheless.

The ROOTS coalition had hoped to expedite the NEPA process because as the timeline lengthened, so too would the amount of input and the pressure on the coalition, which they knew would increase controversy, erode consensus, and ultimately derail reintroduction. However, a lack of continued political and grassroots backing forced its members to spend most of the closing years of the century expending significant time and energy fighting to secure approval for their proposal. Support increased, but so too did dissent, and when the USFWS eventually released the draft environmental impact statement, the issue generated reactions from Americans, regionally and nationally, that were indicative of the divided state of environmental ethics in America at the end of the twentieth century.

One of the main reasons wolf reintroduction had taken so long to implement was because the plan lacked the necessary political support to drive it forward. So in spite of persistent bickering with their environmental colleagues, the ROOTS coalition knew that its most important task was to build political support before the release of the draft EIS, especially in Idaho. At the very least, the coalition wanted to make sure that grizzly bears did not become a conservative punching bag during the 1996 election cycle. The ROOTS plan had enjoyed some initial support from Democratic representatives LaRocco and Williams, but the coalition needed to expand its base of support while continuing to meet the region's diverse political and ideological needs. While finding political support within Idaho and Montana for wolves may have been difficult, a climatic shift in national politics made the chances of achieving consensus on grizzly bears even more difficult.[3]

As a result of the enormous gains made by Republicans in the 1994 midterm elections, the interests and concerns of the Wise Use, County, and Property Rights Movements became a priority on the national agenda. Seventy-three freshman congressional representatives won seats, and Democrats lost their majority in the House of Representatives for the first time since 1954. Speaker of the House Newt Gingrich interpreted the election's results as a mandate to institute a smaller, more conservative federal government and saw rolling back the reforms initiated under the New Deal and President Lyndon B. Johnson's Great Society as his top priority. This meant not only limiting the role of the federal government, championing private property rights, and decreasing spending, but specifically curtailing spending on environmental issues. The Endangered Species Act, which was up for reauthorization, quickly emerged as a favorite target of this new majority, especially in the West. As a result, advocates of the ROOTS plan were fighting to justify the fundamental principle of endangered species recovery on top of the specific details of their plan.[4] Still, the ROOTS coalition hoped its innovative plan would ease conservatives' fears that traditional ESA recovery programs had raised, thus mitigating remaining resistance to the plan and winning support from politicians who opposed a powerful federal government.

They began their campaign with Montana and Idaho's respective governors, Marc Racicot and Phil Batt, who had succeeded Cecil Andrus in January 1995. Racicot, a Republican, was born in Thompson Falls in western Montana and grew up throughout the state, living in Miles City in the east and Libby in the northwest. He stayed in state for college and did not have much of a reputation outside of Montana prior to 2000. He was known for being soft-spoken, humble, and devoid of any pretentions, and these attributes made him a popular figure in Montana. He served as attorney general for four years before being elected governor in 1992. One of his first major accomplishments as governor was the creation of a citizen advisory committee—composed of hunters, outfitters, and landowners from across the state—to address the growing problem of how to manage hunting on Montana's private lands. The group worked together for twelve months on issues such as accessing private lands, protecting habitat, rewarding cooperative landowners, and supporting the outfitting industry. Once the committee submitted its recommendations,

Montana's legislature passed them into law with little debate. This sort of action typified Racicot's penchant and ability for finding creative solutions to difficult natural resource problems.[5]

A few years later, Racicot would testify in front of a U.S. Senate subcommittee on behalf of the Western Governors Association in support of reauthorizing the Endangered Species Act. In his testimony, Racicot not only stressed the importance of consensus, but supported provisions that increased funding to endangered species programs and designated critical habitat for species early in the planning stage. The ROOTS proposal was just the type of innovative plan Racicot welcomed, and once the coalition briefed him on their proposal, he quickly pledged his support. "I'm tired of sitting back and whining and not having any authority," offered Racicot while speaking at a public meeting in Hamilton. "I want Montanans to control it."[6] Racicot even went as far as to lend the services of his attorney to work with Tom France to draft the language and refine the recovery plan over the course of the project. This move signaled that his commitment to making grizzly reintroduction work was genuine and not just an attempt to win political points. Racicot held Montana's governorship until 2001, and although he regularly leveled hard-hitting, challenging questions regarding Bitterroot grizzlies, his criticisms remained constructive, and he consistently worked toward finding a solution.

Montana tended to be more progressive than Idaho, and in the same ways that Montana differed from Idaho, Racicot differed from Idaho's governor, Phil Batt. In 1995 Batt succeeded Cecil Andrus, whose opposition to wolf reintroduction had bordered on militant. Wolves returned to Idaho just days after Batt took office, but Batt was sympathetic to the Wise Use Movement so opposing grizzly bears became the perfect opportunity for him to resume the fight against environmental reform and the federal government, which he believed was violating Idaho's sovereignty. At an initial meeting between Batt, Seth Diamond, and Dan Johnson in May 1995, Batt seemed intrigued by the idea of citizen management and hinged his support on this provision. But by October, his faith in the power of citizen manage-ment had waned. Believing that grizzly reintroduction would create many more complications than wolf reintroduction, Batt noted, "My fear is that if you reintroduce the bears—even in the primitive area—and they start moving out of there, we're going to see the idling of forest product recovery, grazing, the other things which we've been using the land for, to protect the grizzly bear." More simply, Batt con-firmed, "I'm not in favor of reintroducing them."[7]

Batt was not the only Idaho politician to undergo a change of heart that year. Freshman representative Helen Chenoweth, who defeated LaRocco in 1994 and was part of the 1994 conservative sweep, expressed some initial interest in the nonessential, experi-mental designation and the citizen management approach when she met with Chris Servheen in the spring of 1995. By the fall, however, her position on the issue had undergone a complete reversal, and she would litter the press with some of the more colorful and incendi-ary quotes over the next few years. She had already received flak for reacting to the possible extinction of salmon in Idaho by saying they would be available "in a can," and by hosting "endangered salmon bakes," but she had no intention of backing down. Most memorably, she confidently proclaimed, "Artificially introducing grizzly bears to areas traditionally used for human activities makes about as much sense as introducing sharks to the beach." She later added, "The only practical purpose the grizzlies would serve is if they ate gray wolves." Some of her supporters questioned the prudence of this rhetoric, but she never slowed down. She feared for Idahoans' per-sonal safety and criticized the cost of the project—a favorite charge by the new Republican majority—and her provocative rhetoric over-shadowed the more substantive criticisms she occasionally leveled. She was a lightning rod, but her inflammatory remarks were effec-tive, and at many points over the course of the project, they helped energize grizzly bears' opponents.[8]

Senator Dirk Kempthorne was another Idaho politician whom the ROOTS coalition lobbied for assistance, but the first-term senator showed no indication of supporting its plan either. By the end of the decade, Kempthorne would be Idaho's governor, and while he and Governor Racicot shared many surface-level similarities, their differences were subtle but noteworthy. Dirk Kempthorne got his start in government as an executive assistant to the director of Idaho's Department of Public Lands from 1976 to 1978, and he served in a few other roles before being elected mayor of Boise in 1986. Kempthorne held that position for seven years, during which time he built a reputation for finding compromise. His speeches often championed the protection of public lands and the environment, but he rarely took any hard stances on behalf of the environment. While he often sought a balance between prodevelopment and proenvironmental interests, many of his compromises tended to favor development, and during his time in the Senate, he voted with the League of Conservation Voters only 1 percent of the time. His critics constantly questioned his dedication to environmental protection, but the one issue on which no one could question Kempthorne's consistency was his belief in local and state control. He was a committed member of the Wise Use Movement, and his chief legislative accomplishment in the Senate was a bill that prevented the federal government from imposing unfunded mandates on local communities.[9]

Not surprisingly, his interests were very much in line with the Old West. In his first year in the Senate, he sponsored one bill that would have limited the power of the federal government and another that tried to weaken the Endangered Species Act so that destruction of habitat would not be considered a "harm" to species. His proposed amendment to the ESA also would have required wildlife managers to weigh the impact a species' recovery would have on property values before moving forward with any plan, which was a typical concern of the Wise Use and Property Rights Movements. Over the next few years, Kempthorne earned a reputation for compromise and innovation, and because local control was a central demand of the Wise Use Movement, at times he considered supporting the ROOTS plan.[10] But this aspect of his political personality was partisan pandering, and his loyalty to the Old West never wavered. By the year 2000, he would be reintroduction's most powerful and ardent enemy.

Both Kempthorne and Racicot were Republicans, but their positions on Bitterroot grizzlies differed greatly. In this way, these two men mirrored the differences of the states they represented. Whereas Kempthorne was a true conservative who understood the power of consensus strictly as an expedient political tool, Marc Racicot's moderation came more naturally, and his reputation for bipartisanship was sincere. Not only did he tackle issues surrounding hunting on private land as one of his first major acts as governor, but as his support for Bitterroot grizzlies demonstrated, he wasn't afraid to take a controversial stance on an issue if he believed it was right. On multiple occasions, he spoke to groups of resource users on the need to make room for endangered species recovery and for the state to take a leading role in these programs. He was straightforward and honest in his dealings. As a result of this and other successes, Racicot won his second election by a sixty-point margin. He was loyal to Old West conservatives, but his views on certain issues also appealed to Montana's more liberal, New Western faction. In fact, his moderate stances on matters such as abortion made the national Republican Party wary of him; and when George W. Bush chose Racicot as the party's chairman in 2001, the Human Rights Campaign, a prominent gay rights organization, applauded the appointment.[11] As was required of any western politician at the turn of the century, Racicot maintained ideological ties with the Old West, but much like the state he represented, he also had a progressive streak that reflected Montana's shifting identity.

In this way, Idaho and Montana differed greatly, and despite the entrenched opposition of Idaho's political delegation, the ROOTS coalition hosted a series of meetings throughout Idaho and Montana as a way of assuaging these fears, building grassroots support, and addressing concerns. Many people had complained that the process of reintroducing wolves had barred local citizens from participating, so even though these meetings were not part of the formal

NEPA process, they helped foster goodwill among people who believed they had little control in the face of a powerful federal bureaucracy. Defenders of Wildlife also launched a comprehensive grassroots campaign to build support for their plan from the bottom up. For two summers, in 1995 and 1996, the organization's field representative, Minette Johnson, traveled across Montana and Idaho, attending county fairs, farmers markets, rodeos, powwows, community meetings, and other public gatherings advocating for grizzly bears and the ROOTS plan. The first year, a labor representative from Idaho accompanied her on her summer campaign, and the pair was able to convince hundreds of people they spoke with of the merits of their proposal. At the West Idaho Fair in Boise, where Johnson spent a week nestled between booths for organizations like the John Birch Society and local women selling homemade crafts, she discovered firsthand how contentious the issue was. While most people were receptive to her message, by the end of each night's activities, after hours of imbibing, some fairgoers could not help but unabashedly berate her and the idea of grizzly reintroduction as they stumbled back to their vehicles. Even so, Johnson was able to reach out to hundreds of people over the two summers, and grassroots support for the ROOTS plan swelled.[12]

As a result of this support and the group's continued willingness to find consensus and address concerns, national and regional media outlets lavished the ROOTS coalition's innovative approach with unending praise. An editorial in the *Missoulian* called the spirit of problem-solving among traditional adversaries both "remarkable" and "commendable." Many other reporters filled their columns with positive quotes from members of the ROOTS coalition, such as Seth Diamond's hope that the coalition's approach was a chance to "create a new model for endangered species management." People outside the coalition extolled their efforts as well. A biologist for the USFWS was quoted, calling the plan a "unique and unprecedented situation," and another writer called it "one of the most forward-thinking developments on the threatened species front." Referring to the members of the ROOTS coalition and two previous endangered species controversies, an editorial in Idaho's *Lewiston Morning Tribune* called the plan "so balanced and so fair it makes you wonder what these disparate outfits could have come up with if the jobs of saving northern spotted owls and Pacific salmon had been theirs from the start." Finally, another editorial praised the ROOTS plan for being the answer to the Sagebrush Rebellion's call to increase state sovereignty.[13]

Even though positive feelings and a "we can do anything" spirit abounded, some people simply did not want grizzly bears, and no amount of cooperation was going to change that. In its June newsletter, the Montana Stockgrowers Association opened its announcement of the USFWS's July public scoping sessions with, "Uh-oh: It seems importing Canadian wolves hasn't appeased the government's lust for foreign predators." Once the meetings, which were part of the NEPA process, began, emotions escalated further. The gatherings in Salmon and Challis, Idaho, were particularly heated. Minette Johnson attended these meetings to speak on behalf of the ROOTS plan, but these audiences were not nearly as receptive as the people she had met with earlier that summer. After she spoke in Salmon, the audience booed her, which was a first in Johnson's career, but far from the most jarring incident that unfolded at the hearings. In Challis, thirty virulent opponents of reintroduction spoke before the bears' first advocate reached the podium. When he began to express support for recovery, two men menacingly stomped toward him, yelling in a threatening manner, ready to fight. Two police officers restrained the would-be attackers, and although no physical confrontation resulted, the incident left an unsettling feeling that lingered throughout the remainder of the meeting. At the end of the night, one of the officers even suggested that Johnson might check the lug nuts on her tires before driving home.[14] If she had any question about how passionately people felt about the issue, Johnson left the meeting convinced. While grizzly bears may have seemed to someone outside the region as a relatively trivial matter, in the Northern Rockies they inspired as much passion as any political issue.

The meetings, in Hamilton, Missoula, Helena, and Lewiston were far less hysterical, but people still expressed many concerns. In Hamilton, after stating his certainty that grizzly bears would kill women and children, one local resident contemptuously demanded to know who he should sue once it happened. Some Bitterroot Valley residents in attendance openly threatened to shoot the bears if they were introduced. Others were just as adamant in their opposition, but voiced much more restrained criticisms. Tom Greer, the president of the Western Montana Horse Council, had the future of his business in mind when he said, "I have yet to find a recreational horseback rider who wants bears here." Another opponent from Ravalli County averred, "We are, whether we like it or not, reintroducing the bear on the border of the fastest-growing county in the state. It seems inevitable that this will ensure some problems."

Throughout the meetings, John Weaver calmly listened to reintroduction's opponents and tried to address their questions and ease their fears. Hank Fischer left these meetings encouraged, as he maintained, "I haven't heard anything that's unsolvable." Chris Servheen was not discouraged by the meetings' outcomes either, but he had a much different take on them than Fischer. "These people are extremists. They are not representative of the average Montanan or Idahoan," quipped Servheen in a statement he would come to regret.[15] Servheen's statement enraged many people who believed his condescending remark aptly represented the federal government's opinion of local communities, and they did not like being labeled extremists just because they did not want grizzly bears in their backyard.

Bitterroot Valley resident Claire Kelly was one of the "extremists" to whom Servheen referred. Originally from New York City, Kelly had lived all over the country, working for the federal government. Her husband made frequent hunting trips to Montana, and in 1982, she visited her future home for the first time, backpacking in the Bitterroots. A self-described environmentalist and member of the National Wildlife Federation, Kelly immediately fell in love with the area. When she and her husband retired six years later, they decided to settle at the base of the Bitterroots. She loved everything about Montana's outdoors, but deliberately avoided hiking in areas with grizzly bears, so when she heard about the meeting in Hamilton, she decided to attend. Immediately, she found grizzly supporters' claims to be specious, and she felt they tried too hard to minimize the safety factor. When one supporter claimed that the Bitterroot Mountains would be an obstacle that would keep reintroduced grizzlies from coming down into the valley, she responded in her subtle Southern accent, "This grandmother has been back and forth over those mountains a number of times in several different places. If *I* can do it, those bears can do it." She also felt that slots on the citizen management committee would go to people like Tom France and Hank Fischer in favor of local residents. And ironically, her environmentalist background led her to support the conservation biology alternative over the ROOTS plan. Kelly did not like the idea of grizzlies being placed so close to the expanding population of the Bitterroot Valley, and after attending that first meeting, the cavalier attitude of bear advocates hardened her opposition.[16]

Kelly was just one example, but in further contrast to Servheen's claim, many people who opposed reintroduction employed substantive and reasoned arguments that could not simply be dismissed. A former Forest Service employee opposed reintroduction because he believed the mountains no longer contained enough forage to sustain the bears, which would cause them to come down into the populated valleys where they would be a nuisance and a danger. Additionally, he argued that the program would fail because people would not accept bears, so the money would be better spent recovering established grizzly populations. Finally, one resident of the Bitterroot Valley opposed the plan not because of her own personal views on bears, but because she thought the plan was too risky given the acrimonious political climate, and she feared for the safety of the slowly reproducing bears.[17]

Over the years, many westerners had come to distrust the federal government and believed that no matter what it said it would do, it would inevitably overreach. Some of these detractors pointed to Lincoln County, Montana, in the northwest part of the state—where the USFWS translocated four grizzlies to the Cabinet Mountains earlier in the decade to augment the region's population—as an example of the restrictions that Bitterroot residents could come to expect. The initial plan the USFWS formulated would have introduced eight female grizzlies, four subadults, and four cubs, but after locals protested, the agency amended its plan so that it only introduced four new bears. Bruce Vincent of the Coalition for Balanced Environmental Planning, which led the opposition, said in reference to this revised plan, "They've never listened to us in the past. This marks some of the first steps in the right direction." Despite this hesitant optimism, a few years after the bears hit the ground, one Lincoln County commissioner called the augmentation a "major hassle" and complained of the minor restrictions that accompanied the new bears. Serveheen insisted his agency had been responsive to locals' needs, but many Lincoln County residents warned their Bitterroot Valley counterparts not to trust Serveheen or the Fish and Wildlife Service. Some of these charges would have arisen in any case, but whenever an unforeseen inconvenience arose, opponents of the translocation had a new reason to complain. The citizen management committee was a direct attempt on the part of the ROOTS coalition to mitigate the fear that the federal government would impose unwanted restrictions, but many opponents insisted that the committee was just a ploy and unnecessary constraints would inevitably follow a new population of grizzlies.[18]

Despite some negative feedback, support for ROOTS's innovative plan continued to roll in from a variety of sources. In July the results from a survey showed that 62 percent of locals, 73 percent of regional residents, and 77 percent of Americans supported reintroducing grizzly bears into the Bitterroots, while just 26 percent of locals, 10 percent of regional residents, and 8 percent of Americans opposed the action. Additionally, Governor Racicot reconfirmed his support for the ROOTS plan, and he began meeting with groups across the state promoting it. Speaking to the Montana Wood Products Association, Racicot stated, "We have a stewardship responsibility to our native wildlife"; and for him, the ROOTS proposal was the best way to balance this duty with the need to protect industry and jobs in Montana. Furthermore, Montana's senator Max Baucus, who had been unwilling to lift a finger for wolf reintroduction, wrote a letter to Bruce Babbitt encouraging him to back the ROOTS proposal because of the broad and diverse spectrum of support behind it.[19]

The meetings to address the scoping of issues and alternatives concluded in July 1995 on schedule, and the ROOTS coalition hoped the goodwill that the proposal of a citizen management committee had generated among politicians and the media would keep the process on its proposed timeline. But as Batt, Chenoweth, and Montana's other senator, Conrad Burns, soured to the idea of reintroduction, the coalition needed to refine its proposal further in order to mollify their concerns. By the end of 1995, the revised plan allowed the USFWS to issue permits to livestock owners that would allow them to harass bears in a nonlethal manner to discourage the killing and harming of livestock. Along with tolerating the killing of bears in self-defense, the revised plan allowed permitted livestock owners to kill bears found in the act of attacking livestock when federal or state wildlife managers were unable to do so. In response to specific criticisms made by Idaho's Grizzly Bear Management Oversight Committee, the updated plan assured that the citizen management committee would be politically balanced, guaranteed that grizzlies would not bring restrictions on hunting, especially black-bear hunting, and ensured that no trail closures would result. Responding to some of Governor Racicot's criticisms, the new plan also confirmed that the USFWS or Montana's Fish, Wildlife, and Parks Department would remove any bears that wandered into the Bitterroot Valley; and mining, grazing, and recreation would not be considered "harms" to grizzly bear habitat. Finally, the plan tentatively marked recovery at 280 bears.[20]

Even though the group made these revisions, it knew that the plan was far from settled; but this flexibility helped strengthen its base of support. By the end of 1995, the Clearwater Resource Coalition, a local nonprofit dedicated to conserving natural resources in central Idaho, joined the ROOTS coalition; Idaho Outfitters and Guides Association pledged their support; the Idaho Grizzly Bear Management Oversight Committee reconfirmed its endorsement; and Marc Racicot continued to lobby for it. Additionally, Bud Moore, the legendary Forest Service ranger and prominent grizzly advocate, announced his support for the plan.[21]

Still, many other groups lined up against this growing coalition with little likelihood of being persuaded to alter their position. The Bitter Root Back Country Horsemen came out against reintroduction; and in the Bitterroot Valley, an organization called Concerned About Grizzlies (CAG) formed in reaction to the proposed reintroduction. Serving as an umbrella organization for a number of local businesses and nonprofit organizations—including the Western Montana Stockgrowers Association, Montana Wool Growers Association, and Bitterroot Hunters and Anglers—the CAG circulated an antigrizzly petition throughout the Bitterroot Valley that received twenty-seven hundred signatures. The Concerned About Grizzlies group also commissioned a poll gauging attitudes toward bears, and their results indicated that Bitterroot residents opposed reintroduction three to one. Racicot held a meeting in Hamilton that September, but even his assurances were not enough to win support, and the crowd consistently booed anyone who spoke on behalf of the bear, including Racicot. That fall, the Idaho Department of Fish and Game announced their opposition, while still demanding control if reintroduction eventually did happen. A few months later, in February 1996, Idaho's legislature passed House Joint Memorial no. 6, which asked Congress to withdraw funding from the EIS process because of the loss of life, land-use restrictions, and infringement on private rights that grizzlies would bring. Additionally, the bill gave Governor Batt carte blanche to stop reintroduction; and in January 1996, Batt spoke with Rep. Helen Chenoweth and requested she do everything in her power to halt reintroduction from the federal level.[22] Nothing came of this bill, or Batt's plans with Chenoweth, but they established the tone for the remainder of the process that nothing was off-limits.

Although convincing either Batt or Chenoweth to see the merits of the project was a long-shot, ROOTS members convinced Chenoweth to moderate her opposition after meeting with her later that year. Seth Diamond convinced Governor Batt to do the same, but when a summer editorial accused Batt of flip-flopping on the issue, he quickly reaffirmed his opposition, insisting that grizzlies would threaten Idahoans' personal and economic safety. This reassertion of his disapproval served as a powerful reminder that Old West interests had a powerful hold over the region's politicians and gaining their support was a delicate prospect. For most politicians, maintaining a firm ideological connection with the Old West trumped all other concerns. Without this association, their political capital could deteriorate quickly.

Another Idaho representative, Republican Mike Crapo, enthusiastically supported the citizen management approach, but Fischer and France needed someone with more political muscle than the first-term representative. Senator Kempthorne was also in his first term, but he had served as Boise's mayor for a number of years before rising to federal office and held significant political clout within Idaho. Furthermore, he had already emerged as a prominent force in endangered species debates and had shown some desire to promote balanced solutions. Even though he regularly lobbied for Old West industries, at a meeting in Washington, Kempthorne expressed interest in the concept. He did not have any desire to endorse the plan publicly, but he wanted to wait and see if the group could broaden its base of support. Senator Larry Craig was the Idaho delegation's most ardent critic and leaned toward opposing it, but even he admitted that he could get behind the plan if the USFWS adopted it.[23] In retrospect, these equivocations seem specious, but for the ROOTS coalition, relative silence from politicians who would otherwise be opposed was better than the alternative.

In Montana, support for the plan remained more genuine. Racicot's approval helped to dampen opposition across the state, and Baucus's continued support gave the ROOTS plan a boost in Washington. Meanwhile, Montana's other senator, Conrad Burns, considered endorsing the project, but at the very least, promised not to fight it publicly.[24] The conservative political backing that had evaded Fischer and France during wolf reintroduction seemed to be slowly coalescing, or at least remaining neutral. But as Batt's sudden backpedaling demonstrated, the issue was a delicate one and when pushed, even slightly, many of these politicians would quickly revert to conservation's politically safe ideology. The citizen management concept intrigued them, but compromise and consensus were barely emergent theories, while the proeconomic concepts of conservation had been entrenched for decades.

Even though Idaho and Montana's politicians were largely taking a "wait and see" approach to the reintroduction, the EIS was almost a year behind in its original timeline. When the USFWS contemplated not completing it at the end of 1995, ROOTS's proponents feared this hesitancy would mean delaying the sanitation campaign to bear-proof the region's garbage disposal system as well as delaying the launch of a public education program designed to instruct Bitterroot area residents how to live safely near bears.[25]

Education had been a cornerstone of the ROOTS proposal from the beginning. As Steve Nadeau stated, "The more people know, the less they have to fear." This education program would include teaching people the precautions to take while recreating in grizzly country, how to distinguish between black bears and grizzlies, and how to avoid an attack if encountering a grizzly. Because many people had expressed concerns about public safety, both the ROOTS coalition and CBA advocates considered a comprehensive education program as absolutely necessary, and they wanted to begin the program as soon as possible in order to mitigate fears and win approval. They viewed a sanitation program as an essential way to keep negative grizzly-human conflicts at a minimum. Garbage attracted bears, which also increased human contact, and studies had shown this was a leading cause of death for bears. Defenders of Wildlife received two thousand dollars from the International Grizzly Foundation to conduct a sanitation survey that year, and the other members of the ROOTS coalition donated the remaining eight thousand dollars, but these programs would be meaningless if the Fish and Wildlife Service did not proceed with the EIS.[26]

Babbitt again urged the USFWS to complete the EIS, and although resistance on the part of New West environmentalists was largely responsible for the agency's hesitancy, Old West conservationists were just as willing to take advantage of this delay. In its first session of 1997, Idaho's legislature passed a resolution supporting Governor Batt's request to halt the EIS process. The Idaho Fish and Game Commission voted six to one against reintroduction, and even though the USFWS and IGBC met with Governor Batt and other Idaho state officials on five separate occasions that spring, they would not budge. Moreover, a new poll showed that only 43 percent of Idahoans favored reintroduction, while 52 percent opposed it. With new factors creating a more hostile environment for grizzly bears, Senator Kempthorne began to question reintroduction's scientific merits publicly, and soon Crapo turned against it along with Senator Burns, who called citizen management a mask for federal power. In May, Kempthorne requested seventy-five thousand dollars from Congress for a study to confirm that grizzlies had already been recovered across the country and did not face the possibility of extinction, thus negating the need for Bitterroot reintroduction. For all these politicians who were enticed by the citizen management concept, their curiosity was fleeting as they quickly realized that, citizen management or not, a new population of grizzly bears fundamentally challenged their broader interests.[27] As this turn in the tides demonstrated, politicians in Montana and Idaho were merely biding their time until it was politically feasible to oppose the plan and they could do so in a way that did not call into question their commitment to consensus and coalition building.

Kempthorne's position was still officially neutral, but this thinly veiled deception was becoming increasingly obvious. He hoped this study would further delay the EIS process, as the interruption would be a backhanded way of undermining the project's chances for success. Although he was able to accommodate ROOTS's proponents for a time and pay homage to the idea of consensus, Kempthorne's need to appease his Old West constituents in Idaho ultimately dictated his actions in Washington. He had established a reputation for consensus, but as both mayor and senator, Kempthorne consistently proved faithful to extractive industries and his loyalty to Old West interests trumped his willingness to find compromise. He had fiercely opposed wolf reintroduction, had advocated for oil exploration in the Arctic National Wildlife Refuge, and in a few years, he would advocate commencing mining operations in an Idaho state park.[28] Without a doubt, Kempthorne identified with the Old West, and his neutrality began to seem more and more like a mask for his true position.

A shifting political climate allowed Kempthorne, Crapo, and Burns to turn against the proposal, but the ROOTS plan continued to gain national support nonetheless. In the winter of 1996, another poll showed that 62 percent of locals supported reintroduction and only 26 percent opposed it. The following spring, the same polling agency conducted a second survey that showed that the number of locals who supported reintroduction had dropped to 46 percent and opposition had increased to 35 percent. However, this study also revealed that when asked about a plan that kept bears out of the Bitterroot Valley, allowed for citizen management, controlled costs, and did not bring added regulation—essentially describing the ROOTS plan— the number of supporters jumped back up to 62 percent. The hostility many people displayed toward reintroduction was not about the bears. Rather, it was about the presence of the federal government, federal land regulations, and cost. Furthermore, national support for the project remained high. The winter survey found that 77 percent of national residents favored reintroduction. Newspapers in Arizona, Oregon, and Alaska, in addition to the *New York Times* and *Washington Post*, all published editorials praising the groundbreaking nature of the ROOTS plan, and ABC News ran a story that portrayed the plan positively.[29] Even with the intransigence of Idaho's politicians, the ROOTS group had reason to be encouraged.

While much of the rhetoric in the debate was purely political in nature, the opinions expressed ranged across a wide ethical spectrum. Although every comment derived from some ethical foundation, they originated from different eras and philosophical bases. Men like John Muir and Henry David Thoreau had espoused an ethic for the natural world that emerged again in the form of the environmental movement in the 1960s, but conservation was the first environmental ethic to gain appeal on a national scale and to dictate federal policy. By the 1960s, it had gone out of fashion, and environmentalism had replaced it as a more far-reaching, all-encompassing ethic. Even though environmentalism became the dominant ethic for nearly two decades, conservation lingered in the American mind and resurfaced at the end of the 1970s with the coming of the Sagebrush Rebellion. The rebellion fizzled, but conservation remained, so by the time Idahoans and Montanans were seriously debating the merits of reintroducing grizzly bears to the Bitterroots, environmentalism and conservationism were competing for control over the country's natural resource management policy yet again.

Despite the resurgence of conservation, few environmentalists wavered in their beliefs, and those who supported the conservation biology alternative made arguments for the plan that were grounded in archetypal environmental rhetoric. When Matthew Reid of the Great Bear Foundation disparaged the idea of citizen management, he needed only to claim that "science is going out the window." Relying on one of the other tenets of the environmental movement, famed bear biologist John Craighead criticized the ROOTS plan for not protecting the bears' habitat and declared that "all of the high-quality habitat still available in the Northern Rockies should be reserved in the form of congressionally designated wilderness," in order to maintain biodiversity. Employing the spiritual and emotional appeals that were cornerstones of environmentalism, another comment claimed, "Grizzlies are emblematic of the western American wilderness, and of our deep genetic longings for our wild heritage." The survey released in the summer of 1995, cited earlier, asked respondents their main reasons for supporting grizzly bear recovery; the top three answers included saving bears from extinction, saving bears because they were part of the ecosystem, and saving bears because they were here before people.[30] These three responses all exemplified the moral and scientific rationale that drove the environmental movement and demonstrated that environmentalism directly inspired much of the rhetoric supporting bears.

Similarly, many of the arguments made by people who did not want grizzly bears to return to the Bitterroots were grounded in conventional conservation logic. Explicitly invoking the dichotomy between the two ideologies, Rita Carlson, of the Blue Ribbon Coalition, asserted in congressional testimony, "Secretary Babbitt, the Sierra Club, and Earth First do not represent the environmental conscious of this country. We shouldn't call the Green Advocacy Groups environmentalists and passively allow them to refer to us as anti-environmental." Carlson was not only suggesting that conservation was a legitimate environmental ethic that was more representative of the nation's environmental ethos, but was reinforcing that the divide between the two ideologies was stark. Referring to the economic impact that grizzly bears might have, one man commented, "As a fishing outfitter, would I get more business if someone saw a grizzly bear? Probably not." According to another, "These are carnivorous animals that look upon us as food and we are spending millions of dollars to do this. It's a waste of money." By referring to the negative economic impact that grizzlies could have, both statements explicitly invoked conservation's probusiness ideology. Referring to the pro-private-property mantra that had become a central tenet of the Wise Use Movement, another rancher asserted, "I would not watch a grizzly destroy my property or destroy my livestock. I'd take action."

The 1995 survey concerning attitudes toward reintroduction also asked people why they opposed recovery. The leading responses were because grizzlies were dangerous to humans and because grizzly bears had no place being in the Bitterroot Mountains. Confirming this sentiment, Montana resident Michael Koeppen complained that most of the state's major mountain ranges already had grizzly bears and he believed the Bitterroots should be kept free of grizzlies to accommodate recreationists who did not want to deal with them.[31] All of these comments found their ideological origins in conservation's utilitarian philosophies, as Gifford Pinchot had laid them out, and demonstrated how dogmatic the debate had become.

While it is easy to label all grizzly bear supporters as "environmentalists" and all opponents as "conservationists," these labels are slightly simplistic. Most of the people on both sides of the issue did fall into those two prescribed categories; but for others, like Bitterroot Valley resident Claire Kelly, who was an active member of Concerned About Grizzlies, her opposition to the ROOTS plan did not reflect a general adherence to utilitarian, prodevelopment theories of public land management. Kelly openly and genuinely appreciated the need for roadless areas, and she supported the designation of more areas as wilderness; but for her, the grizzly issue was unique. She generally supported endangered species restoration, but in this case, she felt that reintroducing grizzly bears so close to relatively dense population areas was a recipe for disaster.[32]

The main goal of the ROOTS coalition and their innovative plan to reintroduce grizzlies was to forge a compromise between the conservative tenets of conservationism and the more progressive principles of environmentalism, but this created an awkward middle ground not yet backed by a codified ideology. And for as many people who understood the importance and necessity of what ROOTS was attempting, many others were just as determined to stand stubbornly by their ideology. Some magnanimous members of the natural resource community had begun to understand the need to work toward consensus, but on an issue as monumental as grizzly bears, many others believed the matter was too important to compromise and refused to step away from their long-established principles for the sake of balance and cooperation.

In July 1997, the U.S. Fish and Wildlife Service finally released the long-awaited draft EIS, and as most people already knew, it deemed the ROOTS plan—dubbed the citizen management alternative—as its preferred alternative. The draft analyzed three additional alternatives as well. Alternative 2 called for no action and would let grizzly bears repopulate the region entirely on their own. The USFWS rejected this option because studies had already concluded that this approach was not practicable and would take far too long. Alternative 3 advocated an approach that would actively keep grizzlies out of the Bitterroots, and the USFWS discarded it because it contradicted the intent of the Endangered Species Act and the Grizzly Bear Recovery Plan. The EIS also included the conservation biology alternative as Alternative 4, but decided that it went beyond what was needed to achieve recovery.[33]

In addition to laying out the details of the alternatives, the EIS also included cost-benefit analyses. It calculated that under Alternative 1, recovery would take from fifty to one hundred and ten years to achieve. It listed the cost of the five-year reintroduction at $1.9 million and allocated roughly $168,000 for each year after the first five. It forecasted yearly cattle and sheep losses to be around six and twenty-two, respectively. However, it also predicted an economic influx of some forty to sixty million dollars for the region as a result of the reintroduction. The predictions for Alternative 4 were similar, but the draft anticipated more than triple the livestock losses, and contrary to Bader and the AWR's original plan, the EIS foreshadowed a potential loss of up to eleven hundred jobs due to the added restrictions on logging.[34]

Once the draft reached the public, the comment period began. Similar to the scoping process held two years earlier, the USFWS scheduled meetings in Challis, Lewiston, Boise, and Salmon, Idaho, and Hamilton, Missoula, and Helena, Montana, in addition to allowing the public to mail in comments. Controversy and hysteria characterized every meeting, but the meeting in Hamilton was especially contentious. Concerned About Grizzlies rallied their supporters and packed the relatively small meeting house. Not to be outdone, bear advocates, including many students from the University of Montana, traveled south to Hamilton to make their presence felt. Outside the Bitterroot Building, students marched with signs and spoke among themselves, downplaying any safety concerns. Bitterroot Valley residents walked past them and into the building, shaking their heads at their altruistic, yet naïve opponents. Wanting to ease tensions and mitigate any possibility the meeting could turn violent, Chris Servheen organized a private gathering with a handful of Bitterroot residents prior to the meeting to see if there some way to find a middle ground. However, by the time the two sides sat down and agreed on some logistical and administrative parameters, the official meeting was scheduled to begin, and they never had the opportunity for a meaningful conversation. The meeting proceeded as intended, and although the rhetoric escalated, the night concluded peacefully.

The release of the statement created a buzz that resonated throughout Montana and Idaho, and in no time, soaring rhetoric and scathing attacks filled the pages of the region's newspapers. Many of the comments made by ROOTS supporters promoted the merits of the plan's innovative approach. Seth Diamond admitted, "This is an experiment and we don't know if it will work or not, but we are willing to give it a try." ROOTS advocates directed most of their attacks at the bear's opponents rather than their environmental colleagues. "It's highly likely that the bears will be back, and I'm sure the community would prefer that if it's going to happen, that they have some say in what is going on," commented Missoula 's regional forester Hal Salwasser. One editorial scolded Idaho's politicians: "If grizzlies are forced on Idaho, too, the state's loggers, miners, and ranchers need to have their elected leaders tailor the program—not sit on the sidelines pouting." Another advocate, Phil Church, asserted, "You cannot sit there and just say not only 'No,' but 'Hell no.' It just doesn't work that way in today's day and age." A Montana outfitter who worked in the Northern Continental Divided ecosystem remarked, "I find it fascinating that more rhetoric about people not tolerating grizzly bears is heard from people dwelling where there are no grizzlies than from people where the great beasts occasionally roam." Finally, Hank Fischer added, "This ought to be a politician's dream. Look at their speeches: All they talk about is collaboration and working together. Here we have people working together for years and they've handed it to them on a plate and they are still not willing to come forward." Fischer believed the ROOTS group had taken the necessary steps to correct the controversial aspects that had plagued wolf reintroduction, but with this comment, he could begin to see that politicians' complaints ran deeper than they would admit.[35]

ROOTS proponents also reserved some criticism for their environmental colleagues. They no longer wanted to stick devotedly by the environmental movement's ideology only to fail because they did not account for politics. They realized that times had changed since the 1960s and 1970s when environmentalism enjoyed a national mandate, and they knew they had to change with it. "For me the question is, are we going to be idealistic or do we want bears back on the ground?" chided Minette Johnson. Similarly, Defenders of Wildlife's president Rodger Schlickeisen remarked, "Is our goal to make a statement, or is it to make a difference?"[36]

The proponents of a conservation biology approach (Alternative 4) did not waste the opportunity to make their voice heard as well, still refusing to forsake environmentalism's strictest principles. "We feel strongly we need to say what the right answer is rather than bowing to political reality," said Jim Olsen. Another commentator called the citizen management approach (Alternative 1) "short sighted"; and a member of the Greater Yellowstone Coalition added, "We support advisory committees, but we don't want citizens determining the fate of bears." John Craighead's Wildlife-Wildlands Institute added, "To simultaneously impose a socioeconomic experiment designed to dismantle and reinvent the administrative mechanism of endangered species recovery is a recipe for trouble."[37]

From the right, the rhetoric escalated even further. The most creative comment came out of Lemhi County, Idaho, where a community meeting produced the "Lemhi County Anthem," set to the tune of "Home on the Range":

You wanna give me a home
where the grizzlies would roam
and the wild wolves would prowl all day.
Where there won't be around
a job to be found,
and a child would have nowhere to play.

CHORUS
Our home, our home on the range
It's our little home you would change.
And our lost Uncle Sam
just don't give a damn
and he thinks we have nothing to say.

But we know we are right,
and we're willing to fight
To keep the grizzlies at bay.
So our country can see
what it means to be free
It's the good old American way

CHORUS
Our home, our home on the range
It's our little home you would change.
And Old Uncle Sam
had best give a damn
'Cause we've got a whole lot to say![38]

Less whimsically, but no more constructive, one opponent of rein-troduction compared the federal government forcing grizzly bears on Idaho to rape, while another found similarities between the polls that suggested support in the region for bears to the falsified public opinion polls the Nazis had conducted in Austria in the 1930s. At the meeting in Salmon, Idaho, which drew more than 250 people, one person asserted, referring to the ROOTS plan, "In the old days when someone came into our town with a proposal like this, we had a sim-ple solution, get a rope." Although they were in the minority, some Idahoans who had supported wolf reintroduction opposed the same for grizzlies because of the safety issue. Still, many others were much more concerned with issues of land use. "A small minority of people in the west, most of whom are elitists, seek to lock our people out of our environment by pandering to members of Congress east of the Mississippi River who are sympathetic to their so-called 'green' phi-losophy," fired Idaho state senator Ric Branch. While the president of the Idaho Farm Bureau warned, "They're not so interested in preserv-ing the grizzly bear as they are getting control of more land. Once the bears are back in an area . . . those groups will do everything in their power to make the land untouchable." More simply, one Idahoan admitted, "It isn't a fear about the bear. It's the land use." Finally, one Montana landowner remarked, "As soon as they put a grizzly bear in there, they are going to put up the sign and gate the roads, and say this area is closed and you can't go in here," in an allusion to his distrust of the federal government. Similarly, some opponents, such as Helen Chenoweth, viewed the issue as being about states' rights. In Salmon, a detractor remarked, "Citizen management is a placebo; this is about usurpation of state sovereignty." In this same vein, a man who referred to himself as "Cope" noted, "We're still losing the Civil War."[39]

Largely due to resistance from the left, the Fish and Wildlife Ser-vice took two and half years after the EIS process began to release the draft environmental impact statement. Over that time period, the project had gained a national reputation, and although the creativity and cooperation displayed by the ROOTS coalition was largely responsible for this attention, not every reaction was positive. The coalition's fear that a prolonged timeline for the EIS would increase controversy came to fruition as opposition from both sides increased dramatically. The environmental groups that opposed the ROOTS plan did so because they claimed it did not adequately protect the bears or their habitat. But Defenders of Wildlife and the National Wildlife Federation joined the ROOTS group because they believed that the biologically pure environmentalism of the 1960s and 1970s that relied on federal authority and support from the courts was no longer politically feasible. On the other hand, conservative interests and many of the region's politicians resisted reintroduction on any terms because any reintroduction of a large predator like the grizzly bear fundamentally contradicted their ideology. However, ROOTS and the Intermountain Forest Industries Association worked to reintroduce grizzlies because they had too much to lose if they were not involved in the process, and simply saying "no" was no longer a viable or responsible option.

Even though the origins of this controversy were rooted in ethical differences, much of the debate between conservationists and environmentalists split along Old West–New West lines. As a result, many of the sentiments expressed during the public comment period hinted at a deeper regional divide that transcended ethics and cut to the core of the American West's changing social order. With the law on their side, advocates of the ROOTS proposal had reason to be confident, but as Fischer, France, and others involved with wolf reintroduction knew, nothing could be certain until bears were on the ground.

7

The Divided West

Just forty-five beautiful miles through Montana's Bitterroot Valley separate Hamilton from Missoula, and anyone who finds himself in the region would be wise to make the detour along U.S. 93 to see the Bitterroot Mountains shoot six thousand feet skyward from the valley floor, obscuring the majority of the Selway-Bitterroot Wilderness in the process. Despite the short distance between the towns, a visitor would likely notice some significant differences between them. And although many of these differences remain in place today, while the draft EIS was under debate in late 1997, they were blatant, and the two towns represented starkly divergent images of the American West. Hamilton, which identified most closely with the Old West, prided itself as being a small, rural town as opposed to Missoula, which long ago embraced its station as the region's metropolitan center. Moreover, by the late 1990s, Missoula had emerged as one of the epicenters of the New West.

Anyone diligent enough to attend the U.S. Fish and Wildlife Service's hearings in both Missoula and Hamilton in October 1997 could easily identify the dissimilarities between the two. In Missoula, forty-nine of the fifty-six people who testified at the meeting were in favor of reintroduction, and most attendees supported the biologically pure Alternative 4. At the meeting in Hamilton, the split was more even as twenty-two people favored reintroduction and twenty-three people opposed it. But as the meetings reached the smaller, even more rural towns of Challis and Salmon, Idaho, reintroduction's supporters became even fewer. In this same vein, the more urban areas of Boise, Helena, and Lewiston largely favored reintroduction.[1]

Not only did commentators' stance on the issue reflect their allegiances, but the way they spoke about the issue highlighted the differences between the two ideologies. To support their stance, commentators who opposed reintroduction made appeals to Old West sensibilities. As one Idaho man put it, "Our forefathers killed them off. That should tell you something." Another compared the effect that bears would have on people to the impact that the wholesale destruction of the bison had on Indians—classic Old West imagery. On the other hand, comments from people who favored reintroduction embodied the romantic environmentalism that many in the New West embraced, such as "The presence of the grizzly bear enlivens and ennobles humankind," and "To live with the grizzly is an exercise in humility, an admission that we do not and should not control everything."[2] The hearings made clear that the Old West and New West split on the issue of grizzly bears, but their differences ran much deeper.

By the 1990s, Missoula, with a large university and a population of forty-two thousand, shared more in common with places like Bend, Oregon; Boulder, Colorado; and Flagstaff, Arizona, than it did with neighboring towns in rural Montana and Idaho. Farming, ranching, and logging had been staples of Montana's economy since statehood one hundred years earlier, but, by the 1990s, Missoula relied little on any of these Old West extractive industries. The service industry accounted for more than half of its economy, and agriculture, forestry, fisheries, and mining made up less than 5 percent of Missoula County's entire economy. Additionally, federal agencies like the Forest Service had made Missoula a major hub, and nonprofit environmental organizations like Defenders of Wildlife, the National Wildlife Federation, the Alliance for the Wild Rockies, the Clark Fork Coalition, the Sierra Club, the Nature Conservancy, the Rocky Mountain Elk Foundation, and the marijuana reform lobby group (NORML) had all established themselves in Missoula. Writers like William Kittredge, Richard Hugo, and Norman Maclean all called Missoula home at different points, and their writings romanticized Montana's natural beauty, inspiring out-of-staters to flock to its rivers and mountains seeking serenity and solitude. Some came to visit, but many stayed. With this increased population came an influx of young, educated people from California and points east. By 1990, 87 percent of the county's population had graduated high school, in addition to 33 percent who held a bachelor's degree or higher.[3]

These demographic and economic changes sparked a cultural renaissance and added to the wholesale transformation taking place in Missoula. In 1985 *Northern Lights Magazine*, a literary periodical, opened its doors and joined Mountain Press Publishing in Missoula. The city had been home to the University of Montana for more than a century, but in the 1980s and 1990s it became host to art galleries and museums, an orchestra, theater, live music, National Public Radio, and numerous cultural festivals each summer. Its residents could buy the *New York Times*, drink gourmet coffee, and sample locally crafted microbrews. In the summer, Missoulians and tourists alike fly-fished in one of a number of blue ribbon trout streams in close proximity, and in the winter they had their pick of local ski resorts. Completing this transformation was Robert Redford's 1992 film adaptation of *A River Runs Through It*, which put Missoula on the national map as a place where people could find spiritual fulfillment through the meditative experience of casting a fly rod into boulder-strewn mountain rivers. All of a sudden, Missoula became the place where wealthy Californians and easterners could live out their idyllic, rustic fantasies. The town developed a reputation across the state as being extremely liberal, and many Montanans viewed it suspiciously as being overrun with "dope-smoking hippies."[4] For others, however, Missoula's liberal cultural was one of its most appealing features. To say that 1990s Missoula was a far cry from the small, dusty, timber town of the 1870s and 1880s, originally called Missoula Mills, would be an understatement.

Ravalli County and the town of Hamilton, which sits at the base of the Bitterroots, less than ten miles from the western edge of the Selway-Bitterroot Wilderness, changed over this period as well, but it did not see the cultural shift that Missoula did. Rather, the region's isolation, just as much as its scenery attracted new residents; and to some degree, Old West ideologies entrenched themselves deeper in the closing decades of the twentieth century. Between 1970 and 1990, nearly fifty thousand acres of farmland were taken out of production in Ravalli County, and the number of suburban tracts doubled. Even so, one contemporary observer commented, "Many natives of the Bitterroot remain grounded in the land-based values of the Old West." A substantial proportion of the people who were forced to give up farms or ranches remained in the area, resentful of the changes that had forced them to give up their land, and they adamantly supported Old West values. Additionally, although Hamilton's population increased markedly over this period, much of this influx came from retirees who did not have the same impact on the region's culture as those who migrated to Missoula. Many of these people enjoyed the area's isolation just as much as its scenery, and some were fiercely antigovernment or had relocated to escape the increasing diversity of places like California. Hamilton's rural setting enabled these people to embrace the ideology of the Old West, which celebrated an era in which white men reigned over the land without serious competition from anyone—human or beast.[5]

As a result, the infusion of cultural attractions that accompanied Missoula's boom had bypassed Hamilton, leaving it unadorned with New West amenities and with its Old West identity mostly intact. By 1990, 89 percent of the county's residents still lived in rural areas. Ravalli County's population was also aging. In 1980 the median age was thirty-two, but by 2000, it had reached forty-one, and it was only growing older. Hamilton also remained less educated than Missoula. Seventy-one percent of its residents held a high school diploma and only 18 percent had a bachelor's degree or higher. Extractive industries had declined in Ravalli County, but the percentage of people working in agriculture, forestry, fisheries, and mining remained twice that of Missoula County. And although fewer Ravalli County residents worked in traditional extractive industries than the 35 percent of Clearwater County, Idaho, residents who did so, they retained strong cultural ties to these industries—bonds that locals refused to let wither.[6]

The rise of the New West was not unique to Idaho and Montana. By the 1990s, its influence was sweeping across the entire Rocky Mountain region. The split between the Old West and the New West largely overlapped with the divide between those who identified with conservationism and environmentalism, but their dissimilarities went beyond differences in environmental ethics. People who identified with either ideology also split along economic and political lines, places of birth, consumer and recreation habits, education, and values unconnected to the environment, in addition to many other categories. By the 1990s, the New West was well-established in the region as an ideology and a population, and as wolf reintroduction proved, it already had the necessary political clout to achieve major reforms. But if some people considered the success of wolf reintroduction a sign that the Old West's political stature had diminished, the persistence of the grizzly bear debate was harsh proof that the Old West's influence remained salient. Even though the main economic element that defined the Old West—extractive industry—held a rapidly diminishing segment of the region's economy, most of the West's politicians seemed incapable or unwilling to jump ship. Believing that the rise of the New West meant the region's exceptional character would wither, politicians remained devoted to Old Western interests, which ensured that any victory for the New West would be hard fought.

In addition to challenging the dichotomous nature of Americans' environmental ethics at the end of the twentieth century, the ROOTS plan also attempted to unite the Old West and the New. Fischer and France represented the New West, while their timber collaborators represented the Old West, and together, they hoped to unite their disparate ideologies. Even though their plan earned praise from both sides temporarily, when the proposal finally collapsed in the spring of 2001, with its long-sought finish line in sight, the unyielding nature of the divide between the Old West and the New West ultimately was responsible. The debate between the two sides cut to the core of the region's identity, and no amount of compromise could neutralize that fact. The gap between the two ideologies had been growing for more than thirty years, and by the end of the twentieth century, the debate between these two groups was the most essential question facing the future of the West's economy, politics, and culture. It shaped almost every major development in the region, and grizzly bears were no exception. Previous portions of this narrative briefly touched on the conflict between these two groups, but because tension between them was the defining issue that both united and transcended every conflict in the region, a deeper discussion of the factors that created and influenced the divide is vital to understanding the American West at the turn of the twentieth century and to appreciating why grizzly bear reintroduction ultimately failed.

When settlers first moved west, most people hoped to find gold or farmland that was untainted by generations of agricultural toil. Even before them, mountain men trapped furs to trade back east, but from the beginning, Euro-Americans who ventured west did so in hopes of establishing some kind of extractive economy. From those early days, when settlers killed wolves and bears indiscriminately, extractive industries remained the backbone of the West's economy, and well into the twentieth century, the region's identity was based upon the fact that its Euro-American residents worked the land—logging, ranching, farming, or mining. However, in the post–World War II era, this trend began to change dramatically with a surge in tourism to the region. Tourism had been a part of the West's economy since the end of the nineteenth century, but by the late 1940s, its percentage of the region's economic pie started to grow exponentially.

When Americans took to the roads, they traveled to the West more than any other region because it provided recreational opportunities and dramatic, awe-inspiring landscapes that the rest of the country simply did not possess. In addition to being home to soaring, snow-capped mountains, relatively pristine forests, intact ecosystems, and unique wildlife and desert canyons that challenged the human imagination, the region contained unparalleled tracts of public lands. In other regions of the country, the vast majority of land was held in private hands, but in western states, the federal government owned between 30 and 85 percent of the land. This made the region a popular destination because it afforded Americans recreational opportunities the rest of the country could not offer. So while the environmental movement was undoubtedly national in scope, the West became a major battleground for environmental reform because the region's traditional reliance on extractive industries threatened the public lands and recreational opportunities that had inspired Americans to care about the environment in the first place.[7]

Over this same time period, while Americans were discovering the West's rugged beauty and its unique opportunities for hiking, camping, fishing, boating, and wildlife watching, the country was undergoing a much broader shift concerning upward mobility and access to upper-class status. Before the 1960s, social standing in America depended upon birth and shared more similarities with hereditary European aristocracies than most Americans would have probably cared to admit. Admission to elite schools like Harvard and Yale depended upon lineage almost as much as it did on personal achievement. White, Anglo-Saxon Protestants had the inside track on prestige, and private clubs were the centers of social life. While this Old World culture is most obviously associated with the country's urban centers, predominately on the coasts, in the West it manifested itself through land ownership and the number of generations one's family had lived on the land. Native Americans were, of course, excluded from this paradigm, but otherwise, political and social capital in the West, for its white citizens, was more often based on the length of one's ties to the region than on merit-based criteria. Under the Old West's hegemony, this change would mean that newcomers who moved to the region in their own lifetimes inherently did not have the same credibility as those whose families had been in the West for three or four generations. A connection between the old-money elites of the coasts and multigenerational landowners in the Northern Rockies may not be readily apparent; however, both were arbitrary measures of one's status, which those with claims to these identities used as leverage against any newcomers.

By the 1960s, these trends began to change as status and access to wealth became increasingly a matter of merit. Educational institutions democratized as more people had access to higher education, and admittance became increasingly merit-based. As a result, groups that previously had not had access to these channels began to rise through society's economic and social ranks.[8] Nationwide, these changes took hold relatively quickly and new intellectual elites successfully infiltrated the upper class by the end of the 1960s; but in the West, the older class-based system retained its influence, and New Westerners would not effectively challenge its control until the late 1980s and early 1990s.

Many of these newly branded intellectual elite became adherents of the environmental movement. Equipped with educations to which they previously did not have, they subscribed to emerging bodies of science that necessitated the protection of land, air, water, and ecosystems on a broader scale. Many of these new elites were also influenced by elements of the social revolution of the 1960s, which, in part, drew inspiration from the nineteenth-century transcendental movement. Writers like Henry David Thoreau, Ralph Waldo Emerson, and Walt Whitman encouraged readers to seek spiritual and emotional growth through interactions with the natural world—during their vacation and recreation time as well as in their daily lives. Vacation was no longer just a time to relax at the beach, it was a time to learn and grow. So while tourists may have previously been content to stand at the edge of the Grand Canyon or Yosemite Falls, this new upper class wanted to experience these wonders in direct, authentic ways that were unfamiliar, unique, and challenging. And because

this new upper class viewed the land as a medium for spiritual fulfillment and not a place for work, environmentalism complemented
their other sensibilities. For these people, the beautiful lands they had
come to cherish would only continue to possess existential qualities
in their natural state.[9] Once again, the implications of this mentality
were greater in the West where extractive industries—which viewed
the land in less abstract terms as a resource from which to make a
living—had been a staple of the region for generations.

For environmentalists to realize this vision, however, would require
a change in the nation's natural resource policies and in how Americans regarded the importance of their access to outdoor recreation
and wild places. National parks and national forests had been a part
of the West's landscape since the late nineteenth century, but for the
first half of the twentieth century most national forests were managed to benefit extractive industries, and national parks contained
minimal infrastructure and attracted limited numbers of tourists on
a yearly basis. But as Americans continued to travel westward, some
pockets in the region started to cater to these new markets and position themselves as premiere ecotourist destinations. In the 1950s,
President Dwight D. Eisenhower launched the Mission 66 program,
which sought to upgrade the facilities in the national parks in order
to meet this growing demand, and over the following decade, the
national park system became better equipped to handle their growing crowds.

While the federal government was improving its recreational facilities, the ski industry became the first commercial interest to make
a conscious decision to grow itself in the West and capitalize on the
increasing demand for outdoor recreation. States like Utah, Idaho,
and Colorado drew visitors from across the country to their luxury
resorts and unparalleled slopes, allowing wealthy recreationists the
ability to experience nature and challenge themselves against it.
Skiing fulfilled the need of the new upper-class, environmentalist
elite to interact with the landscape. But this desire to know the land
did not end with skiing. Hiking and fishing had always been popular

pastimes, but their mass popularity continued to grow, and within a few decades, rock climbing, boating, and mountain biking joined the mix of common recreational activities in the West. These industries boomed, and by the end of the 1970s, ecotourism had become essential to the West's economy, on the same level as the extractive industries that had preceded them.[10]

In addition to ecotourism, the West also became a destination for heritage tourism. Across the country and the world, travelers began to seek places essential to human history, as well as natural history, as vacation destinations. By the turn of the century, heritage tourism was the fastest-growing sector of the tourism industry. As one observer noted, "The past, it turns out, isn't dead. It's a tourist attraction."[11] In the West, this meant battlefields like the Little Bighorn and the Big Hole; and ghost towns, Lewis and Clark historical sites, the Oregon Trail, and many other features from the Wild West era became major draws that lured people from across the globe. While heritage tourism was indeed distinct from ecotourism, it also necessitated that the land not be "worked" in the traditional sense. Because a desire for education and academic growth drove heritage tourism, it attracted many of the same people who sought spiritual experiences through intimate interactions with nature and identified with the New West.

While the New West's economy was taking root in the region, the position of Old Western industries was becoming unstable and their future prosperity uncertain. As the region's natural resources were becoming more valuable for the intangible benefits they offered, and as the United States' role in the expanding global economy increased, the market value of these natural resources decreased. In the 1980s, the mining industry went bust, so western towns that had depended upon uranium and coal quickly found themselves floundering. Additionally, as more and more New Westerners entered the region, the new economy expanded beyond ecotourism to incorporate technology and other professional industries. From 1970 to 2000, incomes earned from professional and service industries in the West almost quadrupled, from roughly $250 billion annually to just under $1 trillion. Meanwhile, regional income from agriculture and mining industries remained steady, accounting for less than $50 billion annually. By the late 1990s, agriculture was responsible for only 2 percent of Wyoming's economy, while tourism account for 14 percent. In western Montana, by 1980, only five counties employed more than 35 percent of their populations in mining, logging farming, or ranching; by 1994, only two counties did. Similarly, in Idaho, the number dropped from eleven counties in 1980 to seven counties in 1994.[12]

As a result of these foundational economic changes, a transformation had taken place and entire towns saw their economies completely reborn. Red Lodge, Montana, was founded in the 1880s as a coal-mining town, but it struggled once the mines began closing in the 1920s. By the 1990s, Red Lodge's most valuable asset was its proximity to Yellowstone. In the winter, tourism—especially skiing—dictated Red Lodge's economy, and the town had few if any economic ties to its past. Jackson, Wyoming, had a similar story. The first settlers who trickled into the valley in the 1880s made their living by ranching; but by the 1990s, Jackson's prosperity rested in the fact that it neighbored Grand Teton National Park, the National Elk Refuge, and the Jackson Hole ski resort. Ranching persisted in small pockets across the valley, but beyond lending the town some sense of authenticity, agriculture contributed little to the town's economy. Similarly, by the 1950s, the economy of Moab, Utah, was split between uranium mining and tourism. Mining was far more profitable, but when the uranium market went bust in the 1980s, the town turned fully to tourism. Many Moabites were reluctant to rely on tourism as the town's main economic asset because service-industry jobs paid relatively little compared to mining, but the town quickly became a favored destination for mountain bikers and other outdoor enthusiasts, cementing its reputation as a must-see tourist destination, and Moabites eventually embraced it.[13] While these towns were unique by some measures, their circumstances were not totally unrelated, and many other small towns and cities across the West experienced similar changes to one degree or another.

Signifying the extent of this transformation, in late 1996, the Clinton administration sponsored a conference that brought together federal land managers and tourism advocates across the West in order to find ways to promote tourism in the region. Recognizing its importance to the West's economic future, attendees discussed ways to support tourism so that it could grow sustainably over the ensuing years. At the meeting, even representatives from Nevada admitted that the Sagebrush Rebellion's fight over who controlled the public lands was over and that how these lands would be managed was now the most pressing question. Perhaps most representative of the change in opinions related to how public land should be managed, Secretary of Agriculture Dan Glickman commented, "The most valuable thing taken from the public lands is with a camera."[14]

As western towns catered more and more to the burgeoning tourism economy and their new upper-class patrons, visitation to the region increased, and annual visitors eventually sought ways to establish more permanent connections with the landscapes they had fallen in love with on their annual retreats. Many members of the new upper class started buying homes across the Rocky Mountain West, hoping to enmesh themselves in the spiritually and intellectually enriching experiences they found central to their identity. Many of these houses were initially purchased as second homes, but seasonal visitors quickly turned their new retreats into full-time residences, and from the 1970s onward, the population of the Rocky Mountain West exploded. From 1970 to 1990, Colorado's population grew from 2.2 million to 3.2 million. Over this same period, Wyoming's population grew 36 percent from 330,000 to 450,000. By the 1990s, this migration trend was a few decades old, but it had not abated. States like Idaho, Arizona, Colorado, and Utah saw their populations grow more than 10 percent from 1990 to 1994. During that period, 630,000 Americans from the South, Pacific Coast, Midwest, and Northeast all moved to the Rocky Mountain West. Although these New Westerners migrated from diverse points throughout the country, many of them were part of this new upper class, and they moved to the Rocky

Mountain West for similar reasons. Practically overnight, the New West became an identifiable and relatively cohesive demographic.[15]

A majority of these new residents moved to the West's urban centers like Missoula, Bend, and Boulder, where jobs were available, leaving rural areas less affected by this phenomenon. By the end of the twentieth century, many of the West's urban centers had become strongholds of the New West, islands in a sea that still identified with the Old West. However, a certain portion of New Westerners sought out the region's rural areas as ideal places to relocate because it brought them closest to the aesthetic beauty they desired. Regions that were in close proximity to recreational opportunities saw their populations grow the most. These areas experienced development on a scale they had never seen before and subdivisions replaced ranches and farms on a large scale. The amenities that came to the urban areas did not reach these rural regions to the same degree, so in many ways, they were able to maintain their rural persona. But the demographic changes were undeniable, and even though extractive industries persisted, the influx of New Westerners to pockets of the rural West permanently altered the landscape, physically and culturally.[16]

While most people who migrated to the West over the second half of the twentieth century would be classified as New Westerners—in that they subscribed to the principles of environmentalism and had relocated to the West to take advantage of its recreational opportunities—a certain percentage of the region's new inhabitants identified with a set of values that put them in line with the Old West. Much like the influx that hit Hamilton, conservatives from other parts of the country, especially California, started moving to places where their conservative politics would be more accepted as their home states turned increasingly liberal and racially diverse. Rural regions of states like Montana and Idaho quickly became favored spots for these refugees, and soon, these new immigrants became de facto Old Westerners as they identified traditional New Westerners as the very people they had been trying to escape from in the first place. This population was smaller than that of the traditional New West, but it was large enough to maintain and energize the Old West and lend credence to the idea that Old West ideologues were more than just fading relics of a bygone era.[17]

The rise of the New West most immediately impacted the region's economy, but quickly led to new public land management policies and a demographic shift as well. Next in line was a broader cultural shift that resulted from these earlier changes and affected every segment of society. As members of the new upper class descended upon small western towns and cities, they reshaped them from dusty, remote outposts to thriving cultural centers, and many Old West customs were ignored in the process or otherwise assimilated. In order to compete in the new tourist economy, towns started building new or expanded airports to attract visitors and new residents and to make it easier for them to commute in and out. Instead of being isolated in the country's interior, far from any population centers, places like Bozeman, Montana, Jackson Hole, Wyoming, and Aspen, Colorado, became accessible via commercial jet service. Many New Westerners bought into romantic notions of living in a rural setting and purchased ranches as a way of fulfilling this idyllic dream. But unlike their Old West counterparts, they did not work their ranches in a traditional manner. Rather, they let their land lie fallow and merely enjoyed it for its seclusion and beauty. Some of these New West residents were retirees who likewise did not intend to work at all and had moved to the region solely to immerse themselves in the region's unsullied landscapes, thus transforming the rural West from a place of work to a place of leisure.

Many of the cultural institutions that came to Missoula, such as art galleries and festivals, public radio stations, microbreweries, and live music, proliferated throughout the Rocky Mountain West as well. To accommodate the surge in tourist and recreational traffic, high-end companies like Orvis, Patagonia, and Land Rover established outlets and dealerships across the region. The ski industry, which had embedded itself in pockets across the West, continued to expand and became a staple throughout the region. Colorado and Utah had long solidified their reputations as world-class skiing destinations, but by the mid-1990s, Montana and Idaho contained twenty-six ski resorts even though their combined populations totaled less than two million. Completing this evolution from an economically isolated and stagnant region to a prosperous and growing part of the country, by the 1990s, small western towns and cities started appearing in national magazines on lists extolling their virtues as thriving art communities or "best small towns."[18] By any measure the West had changed, and few could doubt that the burgeoning New West was responsible.

Naturally, many who identified most strongly with the Old West did not readily embrace the wholesale change that was remaking the region they had controlled over the previous century, and they found a number of ways to push back against the New West's rise. First, some tried to deny the extent of the change and refused to recognize that extractive industries had declined so dramatically. "In Montana, agriculture is our Number One industry," remarked the vice president of the Great Falls Chamber of Commerce in 2003. Similarly, former state senator Dale Mahlum noted, "Our heritage is cattle ranching and wheat farms. We'd like to have tourism come up to be with agriculture, but not pass it." Both of these comments suggest that many fighting on behalf of the Old West refused to accept the transformation that had already taken place. Not only was tourism just as significant as agriculture, but in certain regions of the West, it had even surpassed agriculture as the dominant economic sector.[19] Considering that some Old Westerners had not yet accepted the fundamental changes to the region's economy taking place, it's not surprising they were slow to embrace grizzly bears.

In spite of these superficial attempts to connect with the region's past, Old Westerners found it increasingly difficult to accept the ways in which the New West was reshaping the region they had controlled for the previous century. First and most importantly, the New West's economic model posed direct challenges to that of the Old West. Old Westerners who relied on logging, mining, ranching, and farming rightly perceived the rise of this new economy, which mandated pristine landscapes and intact ecosystems, as a direct threat to their prosperity. Their livelihoods depended upon working and exploiting the land in the way that their fathers and grandfathers had done before them, and their appreciation for its natural beauty only went so far. The tensions between the two economic models was natural and only increased as efforts to protect the spotted owl and the reintroduction of wolves forced the Old West to realize it no longer monopolized the region's political and economic interests.

The switch from an extractive-industry economy to one based on service industries, especially ecotourism, was difficult for many people in the West who depended on the region's old economy, but this shift hit states of the Northern Rockies like Wyoming, Idaho, and Montana much harder than it did other western states, and their reaction was especially hostile. Unlike places such as California, Oregon, Washington, Utah, and Colorado, Montana and Idaho never received the influx of federal spending during the New Deal, Second World War, and Cold War that had added an industrial element to the economies of these other states. These developments helped those states transition away from extractive industries and create an intermediary step that eased the transition from an extractive economy to a service economy. Montana and Idaho, however, remained dependent on extractive industries through the first six decades of the twentieth century, so the evolution from a colonial extractive economy directly to a postindustrial service economy with no intervening steps was abrupt. Extractive land users had been the sole interest group in the northern Rocky Mountain West throughout its history, so many Old Westerners had difficulty adjusting to this radical change.[20]

While towns like Red Lodge and Jackson may have transitioned relatively smoothly into a serviced-based tourism economy, the region as a whole was not as enthusiastic or prepared for the shift. Because service-industry jobs were notoriously low paying, many Old Westerners had difficulty accepting that these jobs would be the backbone of the region's new economy. As western historian Patty Limerick noted, "In its ongoing vulnerability to the swings of the American economy, tourism may be an unappealing alternative to mining, logging, ranching, and farming. But what else is there?" This was the "devil's bargain" described by historian Hal Rothman. By bringing in desperately needed outside tourist dollars into the region, Westerners invited the marginalization of their local populations and the homogenization of their local cultures.[21] This fundamental paradox left many in the Old West scrambling to reclaim the good ol' days.

Many of the differences between the two groups were entirely economic and resulted from the clash of two economic models that were often mutually exclusive. However, some of the Old West's resistance and the tension between the two groups was less tangible and was, instead, the result of differing values and inflexible adherence by some westerners to the mystique and mythology of the West. The American West had always been a place of myth and legend that stimulated Americans' sense of nostalgia, and popular culture had historically propagated Old West imagery such as pioneers, cowboys and Indians, and mountain men to the point where it was ingrained in American culture. The New West largely rejected the values ascribed to these symbols, but towns like Jackson and Cody, Wyoming, held gun fights every day during the summer months for the benefit of passing tourists, and few people could resist wanting to play cowboy. Rodeos experienced a revival, and dude ranches thrived. Robert Redford's Sundance catalog filled its pages with memorabilia from the Old West like cowboy boots, calico quilts, and Indian pottery, all of which was revamped for wealthy, nostalgic New West consumers. Beginning in the early 1990s, magazines like the *Big Sky Journal* promoted and glorified the affluent lifestyles of that members of the New West, while at the same time using Old West imagery to do so.[22]

The Old West resented the New West's appropriation of its cultural symbols, and argued that the tourism industry's commodification western imagery and paraphernalia was cheap and disingenuous because it lacked the rugged lifestyle that originally molded it. To them, this adoption was inappropriate because items like cowboy boots and cowboy hats lost their significance and authenticity once in the hands of people not connected to the work that originally shaped and created the items. And while the Old West embraced tourism to a degree, it in no way wanted to see this industry undermine the way of life it still lived. As a result, the Old West was wary of the rise of tourism for the cultural and valued-based transformations it incurred as well as the economic changes.[23]

Finally, because the West had been isolated for so long, many rural and small town westerners fostered a natural distrust of outsiders, especially urbanites; so as small western cities continued to grow and gain more political influence, so too did the suspicions of their long-time, more traditional inhabitants. Some Old Westerners even started to view these places as not the "real" West, but something different. As far as the Old West was concerned, logging, ranching, farming, and mining were the sources of the region's authenticity and exceptionalism that had long set it apart from every other region. These industries enabled westerners to live a "pure" existence, untainted by urbanization, industrialization, and all the corrupting influences of modernity.[24] Even though these industries existed only on the fringes of the West's economy by the late 1990s, people who identified with Old West culture still believed it to be authentic. Never mind population centers like Missoula, Helena, Kalispell, Great Falls, Bozeman, Billings, Boise, Nampa, Meridian, Idaho Falls, or Pocatello—these places were not the *real* West as far as many rural residents were concerned.

Ironically, in the face of the critique that their way of life lacked legitimacy, many New Westerners migrated to the region to reclaim a sense of authenticity and masculinity in reaction to an increasingly urban world. As critics of western literature, William R. Handley and Nathaniel Lewis, write, "There is no other region in America that is as haunted by the elusive appeal, legitimating power, and nostalgic pull of authenticity."[25] However, the two groups defined these traits much differently. Whereas the New West considered submitting to nature in its entirety, which included reintroducing potentially dangerous animals like grizzly bears, as a defining masculine and authentic experience, the Old West believed that subduing and conquering nature was necessary to lay claim to masculinity and ward off modernity's corrupting influence.[26] Adding to the tensions, while many people who identified with the New West moved to the region for its isolation, the mass influx of new residents made that harder and harder to achieve. Throughout the 1990s, places like Montana's Bitterroot Valley saw more and more of its ranches subdivided and turned into planned communities. So there existed a degree of hypocrisy as to how New Westerners were altering the landscape. They were destroying the very thing they sought, and this fact was not lost on many of their Old West counterparts who resented the loss of their open spaces.

By the 1990s the impact of the New West on the region's social landscape was undoubtedly growing. Economically, it had long since surpassed the Old West. Culturally, the region was divided more or less evenly. But politically, the Old West still held a distinct advantage. Because the West's identity was directly tied to nostalgia for a bygone era, claims to authenticity and tradition offered the easiest access to political capital. With the idea that the West's heritage was second only to its scenery as a defining feature of the region, the Old West was the politically safe bet, and politicians were unwilling to deviate from this model or risk supporting a policy as bold and iconoclastic as reintroducing grizzly bears. Even some New Westerners, who worked within the region's modern economy and supported the change that was occurring, kowtowed to the Old West's notions of authenticity and generations on the landscape as legitimate prerequisites to wielding influence within the political realm.[27]

Much like Frederick Jackson Turner's argument in his famous essay "The Significance of the Frontier in American History," Old Westerners feared that if the region lost the economy and culture that had shaped it over the previous century, the West's exceptionalism would wither. Turner argued that the frontier experience—the opportunity for unlimited westward expansion—was the essential experience that had shaped the country's entrepreneurial spirit and unique character. As a result, in the decades after 1890, Americans struggled with what historians have termed "frontier anxiety." Because they no longer had the opportunity to settle and conquer wild lands, Americans believed the country's exceptional character would wilt, men would no longer have the opportunity to prove their masculinity, and Americans would become indistinguishable from Europeans. To combat this descent, turn-of-the-century Americans looked for new outlets to relive the experience of the frontier. The rise of organized sports, the formation of the Boy Scouts, transcontinental expansion, and nostalgic depictions of the frontier era (like Buffalo Bill's Wild West Show) all were attempts to preserve this moment.[28]

By the time environmentalists were trying to reintroduce grizzly bears to the Bitterroots, the West was on the verge of a new frontier—an economic and cultural frontier brought on by the rise of the New West. More and more, Old Westerners worried that the Northern Rockies were becoming similar to its neighboring regions, and they were struck with a similar bout of frontier anxiety. Just as turn-of-the-century Americans tried to preserve elements of frontier culture and recreate that experience, Old Westerners and their faithful political representatives were wedded to the same traditions and pursued policies that sought to preserve what they perceived as the source of the region's exceptional character—its economy. The rapid transformation from relying on extractive industries to service industries such as tourism left many Old Westerners yearning for the rugged lifestyle cultivated by logging, ranching, farming, and mining. Without these, they too believed, the West's exceptionalism would evaporate, and men would lose their masculinity and claims to an authentic existence. Under this theory, westerners would become no different than other urban and suburban Americans whose ways of life they considered inauthentic.

As a result, even though tourism was a cornerstone of the region's economy by the 1990s, and nearly four million people visited Yellowstone each year, politicians from Montana, Idaho, and Wyoming refused to support wolf reintroduction because it threatened the ranching industry. In this same vein, although the USFWS estimated in the draft EIS that grizzly bears could bring an added forty to sixty million dollars to the region, almost every Idaho politician, and many in Montana, refused to see the plan's economic merits.[29] As far as they were concerned, supporting policies that harmed traditional extractive industries was part and parcel of surrendering the Old Western way of life. So as the Bitterroot grizzly EIS process slowly stumbled forward, receiving more and more criticism, most of Montana's and Idaho's politicians, even those who had expressed some level of support for the project, eventually gravitated to the Old West point of view.

Many people were aware of this phenomenon on some level. As one land owner stated, "In all honesty, I don't think this is about grizzly bears. This is about power." While certainly accurate, statements such as this, and the majority of people who made them, disdained the rise of the New West and saw the potential reintroduction of grizzly bears not as a symptom of a changing power dynamic, but as a means to it. By this line of thinking, if grizzly bears returned, all else would be lost. By the end of the twentieth century, however, much of the region's transformation had already been completed. The changes that had swept the Northern Rockies had already brought the region from a colonial, extractive economy to a postindustrial, service economy.[30] With the exception of an occasional mining boom and subsequent bust, the chances that the economy would ever revert back again were slim. So while many people fumed over the idea that grizzlies would mean losing power in the region, they were blind to the fact that the region's economy and culture were already progressing in a certain direction, whether grizzlies returned or not. But because nostalgia and authenticity were forms of currency on par with economic and demographic factors, the Old West was still able to compete successfully against the New West.

By the end of the twentieth century, the clash between the New West and the Old West was the essential dilemma confronting the region. The two sides' differing visions of the West affected every element of society, and while the debate over grizzly reintroduction did not, by any means, invent this impasse, it did bring it to the forefront of the region's consciousness. The ROOTS plan had hoped to forge a compromise in which each side could claim victory, but the divide between the two had become so fundamental and dogmatic in nature that compromise was not possible. The Old West could not accept reintroduction at any price because the return of grizzly bears would mean a victory for the New West. On the other hand, the New West did not want to allow the continued presence of logging in areas where grizzlies would be reintroduced because that would mean the persistence of vital elements of the Old West. Compromise was an alluring and enticing prospect that had mitigated the influence of this dichotomy for the first few years of the project. As it moved forward, however, the multifold layers concealing the essential nature of the debate were slowly stripped away until it became clear that reintroduction was in fact a conflict between competing visions of the Old West and the New West. Once this became apparent, the ROOTS plan could not escape the volatile and capricious nature of the singular debate facing the future of the West at the turn of the twenty-first century. This would be its undoing.

8

Triumph and Collapse

In November 2000, the Fish and Wildlife Service published a Notice of Intent to implement the citizen management approach to implementing grizzly bear reintroduction in the Bitterroots (the plan that had been designated as Alternative 1). But by the following year, reintroduction's prospects looked grim. Even so, Missoula and much of western Montana had "Griz fever." For the second time since the environmental impact statement process had begun, University of Montana Grizzly football was on its way to a national championship. The town was abuzz. Everyone from the retiree waiting in line at the sandwich shop to the young girl working behind the counter covered themselves in Griz paraphernalia. On the radio, a call-in show was bombarded with calls from fans showering Griz football with lavish adulation, and even local weathermen could not resist making references to the team during their broadcasts. And if anyone wanted to find a motel room in town on game day, good luck. In December, when the team traveled to Tennessee for the national championship game, police cruisers with flashing lights, pickup trucks full of cheerleaders, and a fire truck with Monte, the Griz mascot, escorted the players to the airport caravan-style. Even in Hamilton, residents were excited about the Griz's title run. No one put it better than Montana's congressional representative Rick Hill when he commented, "They cheer the grizzlies on Saturdays down there in Ravalli County, and they curse them the rest of the week."[1] Western Montana was wild for grizzlies—that is, as long as they wore helmets, pads, and polyester as opposed to fur and three-inch claws. So just as the opportunity of returning *actual* grizzly bears to Missoula's backyard was fading into the sunset, many western Montanans were too distracted by the triumph of Griz football to notice.

As contradictory as this may have seemed to some observers, it was not altogether surprising. The West had long been a place of myth and symbolism, and Montana was no different. In Texas, the cowboy persisted as the iconic figure; in Arizona, the saguaro cactus emblazoned the state's license plate; as did the bucking bronco in Wyoming; and the Zia sun symbol in New Mexico. Throughout the region, the noble Indian and lone lawman have endured as some of the West's favored myths, although their significance, even during their supposed heydays, was questionable. In Montana, the lasting image was that of the grizzly. Perhaps no animal better symbolized the West than the grizzly, and despite its controversial reputation, Montana proudly claimed it as its state animal.

So naturally, football season was not the only occasion during which Montanans celebrated the grizzly bear. In fact, grizzly imagery had long embedded itself across the state. Down by Missoula's waterfront, in front of the offices for the Boone and Crockett Club, a massive bronze statue of a standing, snarling bear overlooks joggers and dog walkers as they pass. Similar references are plastered about town on store fronts and public buildings. When college students need a quick snack, they stop by Grizzly Grocery, and when most Missoulians are in the mood for a hard drink, they make a trip to Grizzly Liquor. Grizzly Hackle is the city's favorite fly-fishing shop, and when residents want to ride the state's open roads and feel the wind blow through their hair on a motorcycle, most of them wander over to Grizzly Harley-Davidson. Some Missoulians even rent their apartments and houses from Grizzly Property Management. And in case visitors happen to miss any of this iconography, a towering taxidermied bear greets them as they enter the baggage claim area of the Missoula airport.

Although grizzly bear imagery is especially prominent in Missoula, the rest of the state is similarly enamored with the great bear. Bozeman's airport also has a bronze statue of a grizzly to greet passengers, and in other parts of the state, one can patronize businesses with names such as Grizzly Floor and Tile, Grizzly Auto Electric, Grizzly Saddlery, and Grizzly Custom Guns. In 1992 Alaska Northwest Books published a book of essays extolling the majesty of the great bear, which featured contributions from two of Montana's leading literary voices, William Kittredge and A. B. Guthrie Jr. Even in Salmon, Idaho, a place where local citizens passed a resolution stating that grizzlies had never resided there, an enormous bronze statue of a grizzly stands at the center of town.[2]

Despite the fact that Montanans seemed proud of and celebrated this piece of their heritage, these sentiments remained largely superficial. While many Missoulians were indeed in support of reintroducing grizzly bears, the tradition of celebrating the myth and spirit of the grizzly, if not the bear itself, often overshadowed the desire to protect real grizzlies. In this way, the New West's obsession with the myths and symbolic imagery of the Old West was harmful to their greater goals. Whether they realized it or not, Montanans became comfortable with the grizzly inhabiting a strictly symbolic function and were content with this being their primary relationship with the bear. The tension between the Old West and the New West made it even more difficult for the bear to break free of this superficial identity, so even as reintroduction's prospects slowly waned as the result of sociopolitical pressure, the state's relationship to the symbolic grizzly never wavered.

Before the Republican Bush administration took office in 2001 and indefinitely shelved the project, the U.S. Fish and Wildlife Service proceeded slowly, if not without obstacles, following the release of the draft EIS and subsequent comment period. From the right, opponents increased their attacks on the project's funding and stepped up their media campaign, harping on the troubles that grizzly bears would bring to the region. From the left, environmentalists continued to tout the biological merits of the conservation biology alternative, while disparaging the politically compromised citizen management plan (the preferred alternative of the U.S. Fish and Wildlife Service); and when that proved ineffective, they laid the groundwork to trigger their traditional safety net—the courts. Amidst this strife, the Fish and Wildlife Service worked to incorporate as many of the suggestions expressed by the public as it could into the plan in order to solidify popular support while meeting the needs of the bears. By this point, the USFWS had taken the lead on revising the alternatives, but the ROOTS coalition continued to lobby political leaders and leverage their grassroots networks to mitigate the fears people expressed during the comment period. In the spring of 2000, the USFWS released the final EIS, which again chose the citizen management alternative as the preferred alternative, and in the fall it made final preparations to begin reintroducing bears in the summer of 2002 under this plan. After eight years of working together, the ROOTS coalition finally expected to see its hard work pay off, which for people like Hank Fischer, Tom France, and Dan Johnson made its sudden unraveling so devastating.

By the end of the comment period, more than twenty-four thousand people had mailed the USFWS their opinion on the preferred alternative and suggestions to improve it. Although almost 90 percent of the comments were in the form of petitions, individuals wrote letters and mailed postcards that expressed a range of opinions. The majority of respondents were from Idaho and Montana, but people from across the West and the country accounted for a significant number of comments as well. The majority of respondents supported reintroduction, but most favored the conservation biology approach over the citizen management alternative. In total, 1,300 people attended the USFWS's seven hearings, and of the 294 people who spoke at the meetings, 157 supported grizzly recovery, 103 opposed it and 34 were undecided. Again, however, while most speakers wanted to see grizzly bears back in the Bitterroots, the majority favored the conservation biology approach over the USFWS's citizen management plan.[3] This curious outcome was the product of multiple factors that were converging at the end of the twentieth century in the West. On the one hand, the popularity of groups such as the Alliance for the Wild Rockies had swelled, especially among young people, because of its support for the Northern Rockies Ecosystem Protection Act. At the same time, however, wildlife professionals like Hank Fischer and Tom France were much quicker to recognize the need for compromise than were the average supporters of endangered species restoration. The ordinary citizen-supporter of such programs did not have to deal with the daily struggles involved in advocating for animals like wolves and grizzly bears. While most people did understand that wolf reintroduction had been a long, hard-fought battle, they could not really appreciate why Fischer and France believed so strongly in their ROOTS-style approach in the case of grizzly restoration. Even so, these comments reflected the opinions only of people who had felt compelled to write in, which was not necessarily representative of the public at large; but proponents of the conservation biology alternative took these numbers as a positive sign and went on the offensive.

Following the release of the draft EIS and public comment period, popularity surged for the conservation biology alternative. The Nez Perce and Confederated Salish and Kootenai Tribes both came out in favor of this alternative, as did the World Wildlife Fund. Hoping to capitalize on this boon of public support and the issue's increasing divisiveness, Louisa Willcox of Wild Forever urged the USFWS to overhaul the entire plan in favor of the conservation biology strategy. She and other proponents of this approach continued to criticize the experimental designation, the lack of codified habitat protection, the authority and legality of the citizen management committee, and the size of the recovery area, but they also launched new attacks on the logistics of the reintroduction.[4]

The year before, at the prompting of Wild Forever, a group of forty-three Canadian environmental organizations had written an open letter questioning the prospect of grizzly bears being taken from

Canada to be put in the Bitterroots. They claimed that the status of Canada's grizzly population was uncertain, and they wanted assurances that the translocated bears would survive in their new homes. Over the course of the following year, hard-line environmentalists in the United States continued to cite this objection in addition to disparaging the idea of taking grizzlies from either the Greater Yellowstone or Northern Continental Divide ecosystems. They opposed taking bears from the country's two healthiest grizzly populations outside of Alaska, both because they were still threatened and because removal of individual bears could be detrimental to the populations' recovery. On top of that, they resisted this action because individual bears would have their status under the ESA downgraded. By leaving a threatened population and joining an experimental one, the bears would be exposed to threats from which they had previously been protected.[5]

In response to this, Fischer contended that only bears that had moved outside the recovery areas would be translocated to the Bitterroots. And in an attempt to appease their colleagues, Hank Fischer, Tom France, and Sterling Miller (a recent addition to the National Wildlife Federation who had worked as a bear biologist in Alaska) shelved the idea of taking bears from the NCDE because its population was not as healthy as that of the GYE. However, they kept open the possibility of taking bears from the GYE and suggested using bears from Alaska as well. Still, British Columbia, which was home to an estimated ten thousand bears, remained the leading candidate from which to translocate the reintroduced bears, and because British Columbia and Yellowstone contained habitat much more similar to that of the Bitterroots than did Alaska, they deemed translocation from the forty-ninth state a last resort even though it boasted the largest population.[6]

In response to other critiques from the left, advocates of the citizen management alternative argued that the committee would have a scientific voice through the representatives from the state and federal wildlife agencies. They also tried to alleviate fears concerning the clause that permitted citizens to kill bears by asserting that state and federal officials had taken care of every issue in the other ecosystems over the previous twenty years without fail, so the clause would go largely unused and untested. Despite these attempts, environmentalists refused to give any ground, sticking to Mike Bader's earlier sentiment that no reintroduction was better than a compromised one.[7]

Dialogue between these factions collapsed once again, as supporters of citizen management knew their coalition depended on compromises that advocates for the conservation biology alternative found untenable. Additionally, citizen management advocates found the unwillingness of their environmental colleagues to compromise contrary to the spirit of the delicate arrangement they had crafted with the timber industry, and they eventually stopped trying to work with them. Supporters of the ROOTS plan believed the conservation biology approach would be impossible to implement given the hostile political climate and would fuel the persistent accusations that grizzly bears were just a thin veil to lock up land.[8] Knowing that winning acceptance from people who were unsure or outright hostile to grizzly bear reintroduction would be the difference between the plan succeeding and failing, the USFWS and other proponents of the preferred alternative (citizen management) spent most of their limited time, money, and energy addressing these concerns.

The citizen management committee was by far the most innovative piece of the USFWS's preferred alternative and was supposed to be the element that made the idea of grizzly bears acceptable to people who had resisted reintroduction. But it could not shake its doubters. While environmentalists believed the committee was illegal and would be loaded in favor of extractive industries, people who feared federal overreach alleged the committee's authority was merely an illusion and the secretary of the interior would snatch power away from the committee as soon as he or she had the opportunity. Responding to these charges, the ROOTS coalition consulted outside legal advice, which confirmed that the secretary of the interior had the power under the ESA to delegate his or her authority. But to ensure the secretary had ultimate authority, yet not wield it arbitrarily, ROOTS and the USFWS added to the plan a six-month review that the secretary of the interior would be required to institute and complete before taking power away from the citizen management committee. Once this clause made its way into the alternative, environmentalists complained that the process would be too cumbersome and that no interior secretary would want a direct showdown with the governors of Idaho or Montana. Still, the change made its way into the final EIS in addition to guarantees that the committee would be politically balanced and comprised of people who lived near the Bitterroots.[9]

While claims of federal overreach and anxiety about future land-use restrictions inspired the most common complaints from the right, fear for personal safety persisted as a much more tangible issue. For the Bitterroot Valley group Concerned About Grizzlies, fear drove their opposition more than any other factor. This was a matter that bear advocates were ready to tackle. "The issue of public safety is legitimate," conceded Chris Servheen, though he added that "the risk of being killed by a grizzly bear is extremely small." Fischer also noted, "There is a fear of bears that I think is similar to the fear some people have of flying on airplanes." To defend the improbability of an attack, supporters pointed out that grizzlies had killed only one person in the Bob Marshall Wilderness outside of Glacier National Park since 1950. Additionally, bears had killed only nine people in Glacier between 1913 and 1995. In Yellowstone National Park, only five people had died from bear attacks between 1839 and 1994, and only three more had died in the Greater Yellowstone ecosystem in that time period. Moreover, Servheen surmised that bear death statistics from national parks were inflated because national parks received more visitors who tended to be less informed about how to travel safely in grizzly bear country. Wilderness areas, like the Selway-Bitterroot and Bob Marshall, saw far fewer visitors than Glacier, thus making the statistics from the Bob Marshall a better indicator of what could be expected in the Selway-Bitterroot. Furthermore, Servheen insisted that a fully recovered grizzly population in the Bitterroots would be much less dense than the ones in either Yellowstone or Glacier. Another commentator estimated that even after the population recovered, there would be only one death per half a billion visitors. In a final attempt to alleviate personal safety concerns, Servheen added, "In general, the people that live with bears have fewer concerns than those that don't."[10]

In addition to publicizing this array of statistics, bear advocates hosted a handful of meetings featuring outfitters and other people well-versed on how to travel safely in grizzly country in an effort to allay critics' fears. The meetings were well attended and positively received, and advocates were dedicated to continuing this process; but some opponents refused to be convinced that grizzly bears did not pose an immediate and direct threat. For Claire Kelly, these advocates were vastly overselling their position. While she understood that human-grizzly interactions were often benign, she believed the degree to which advocates were downplaying the safety issue was specious and increased her distrust even further. "Let's save the forests. Let's save the animals. But let's tell the truth," quipped Kelly.[11]

For others, however, the fear was much more visceral. In this way, grizzly bears filled the role of the "alpha predator" that is the focus of David Quammen's book, *Monsters of God*. According to Quammen, "Alpha predators and the responses they evoke, have transcended the physical dimension of sheer mortal struggle, finding their way into mythology, art, epic literature, and religion." Reflecting this evaluation, one editorialist maintained that grizzlies were "five-hundred-pound eating machines with teeth like talons and the disposition of Mike Tyson." Openly eschewing the statistics, one Idahoan noted, "It's a known fact that it's a threat to man. I don't care about the percentages, it's dangerous." Referring to the reintroduction, Idaho representative Helen Chenoweth added, "It may only kill one in 10,000 or so, but it is still not a good thing to do." Confirming this sentiment, a final editorialist retorted, "I don't care about statistics. I care about being eaten. Somewhere out there, a grizzly is waiting."[12] These persistent fears may have seemed irrational given the statistics, but they remained very real for some of reintroduction's opponents and made for effective rhetoric that lingered in the public debate.

While many critics could not get past the fear factor, Montana's governor, Marc Racicot, who favored the preferred alternative, continued to prove himself one of the most thoughtful and intelligent reviewers. First, he wanted the USFWS to add a stipulation that any bears taken from the GYE or NCDE would not be counted against that population's annual mortality rate—a key statistic used to gauge a population's long-term sustainability—so that delisting in those areas would not be delayed. Over the objections of environmentalists, the USFWS assured him this would not be the case because any bears extracted from the GYE or NCDE would be ones that resided outside the boundary of the recovery areas and thus would not count toward the recovery total anyway. Racicot also wanted to make sure the citizen management committee would have true authority and that no hunting restrictions, especially, on black bears, would result. The USFWS readily accommodated this request as well. His final request, and the one on which his full support ultimately hinged, was for the USFWS to ensure that the federal government funded the project for its entirety and that money spent on Bitterroot recovery would not detract from funding in Montana's other grizzly bear recovery areas.[13]

If the USFWS had planned on granting Racicot this final request, larger political forces worked to undermine that guarantee. In July, Senator Conrad Burns of Montana and Senator Larry Craig of Idaho introduced a bill that cut the project's funding for 1998 and required a habitat study to confirm how many grizzly bears, if any, the Bitterroot Mountains could support. Bear advocates bristled over the effects this delay would cause, and they believed the habitat study was unnecessary because of the ones conducted a decade earlier. Mike Roy fumed, "The public has been waiting for this EIS for a long time and now the senator wants to shut down the process." In reference to the bill, Senator Craig replied, "It doesn't say there will never be reintroduction. It just says this viability study has to be done first." Although bear advocates tried to fight the rider, President Clinton eventually signed the appropriations bill to which it was attached. Funding for 1998 was cut from $150,000 to $75,000. The stipulation preventing the reintroduction of grizzlies to the Bitterroots that year was mostly symbolic because the final EIS had not yet been released. However, it set a precedent that made bear advocates uncomfortable and was one they were all too familiar with from their efforts with wolf recovery. The USFWS assured Racicot that funding for Bitterroot recovery would not be taken from the money demarcated for recovery in ecosystems, but the agency made no further promises. Racicot maintained his support, but he continued to press the USFWS for answers about funding, few of which they were able to provide.[14]

Although the ROOTS coalition spent most of its time trying to convince doubters of the citizen management alternative's merits, it continued to find ways to improve the plan, and it, too, submitted suggestions to the USFWS. In addition to supporting all of Racicot's recommendations, the coalition wanted the USFWS to form the citizen management committee as soon as possible and put it in control of the education and sanitation campaigns. The coalition believed that education would be essential because as people knew and understood more about grizzlies, they would have less to fear and be more likely to accept them. Also, it stressed the need to sanitize the experimental recovery area and make it "bear-proof" by installing bear-resistant trash cans and waste systems. Human garbage could be a powerful attractant, so this would considerably reduce the chances that bears would wander into populated areas and dramatically decrease the chances of negative human-grizzly interactions. The ROOTS coalition wanted both of these things to happen before bears hit the ground, and the quicker they could accomplish this, the sooner the committee would build good faith and prove to doubters that it could function effectively. Also, the group urged the USFWS to substantiate the claim that reintroduction would bring forty to sixty million dollars into the local economy, believing the dearth of statistics supporting it in the draft EIS called the entire document's credibility into question.[15]

While the USFWS was making final changes to the EIS, Idaho's Senator Kempthorne remained publicly neutral, but began criticizing the possibility of reintroduction saying that the federal government had not adequately included Idaho in the process. In spite of Kempthorne's assertion, the state had had a number of opportunities for input, some of which it had openly shirked. Not only had Idaho attempted to exclude itself with the creation of the oversight committee, but the state had been actively involved since Idaho's Fish and Game director, Wayne Wakkinen, chaired the working group during the writing of the Bitterroot recovery chapter in 1992. On top of that, the preferred alternative gave Idaho seven of the fifteen seats on the citizen management committee. Kempthorne had previously made other subtle attempts to derail reintroduction, but similar to strategies employed by Burns and Craig, Kempthorne's opposition was becoming more and more obvious. Throughout the project, these politicians had felt pressure to support the ROOTS plan because it had a broad base of support and answered their calls for greater local control by giving the states unprecedented management responsibilities. But when push came to shove, the plan fundamentally challenged the hegemony of the Old West, and for these politicians, maintaining the Old West's power ultimately trumped these more recent political trends.[16]

The New West, as a political constituency, was still in its infancy compared to the entrenched values of the Old West, and as controversy increased, few of Montana and Idaho's politicians could resist the comfort and ease of falling back on Old West values, which many in the region still identified as more authentic and credible. Even Democratic senator Max Baucus had withdrawn his unequivocal support; and any chance Kempthorne would alter his position to support reintroduction was rapidly diminishing. By the beginning of 1998, Governor Batt decided not to seek reelection, and Kempthorne quickly emerged as the front-runner to take his place.[17] As he looked forward to returning to Idaho, Kempthorne knew he no longer had to pay deference to bipartisanship in Washington. Facing election in his home state again, he had to appeal to his Idaho constituents who largely adhered to the values of the Old West and conservation.

As Kempthorne prepared to return to Idaho, he felt pressure from the Old West to conform to the strictest form of its ideology, and this was not unusual. Politicians and bureaucrats at all levels of politics in the Northern Rockies were forced to pay homage and abide by its ideals. Few careers exemplified this phenomenon more than that of Steven Mealey. After working as a hunting and river-rafting outfitter on the Salmon River, Mealey attended the University of Idaho where he studied wildlife and forestry. He graduated in 1975, and for the next few years he worked as a grizzly bear research consultant out of Bozeman before accepting a job with the U.S. Forest Service as a wildlife biologist. In a 1982 article in *Audubon*, Mealey cited the effect the rise of the New West would have on grizzly bear recovery: "This is a new era in wildlife management, and there's a cultural change in our society now under way that's going to bring the necessary public support." He continued, "Bears and humans have lived in proximity for hundreds of years. It can be so simple." Over the next decade and a half he held a handful of different positions with the Forest Service, but in 1997 Governor Batt appointed him to be the director of Idaho's Department of Fish and Game. Up until this point, Mealey's career and personal philosophy embodied the spirit of the New West, but soon after he assumed his new position, he took a hard-line stance against reintroduction that eschewed facts and promoted misinformation. This abrupt ideological shift was a reaction to Idaho's political climate and a betrayal of his career up to that point. Not only did he oppose the possibility of reintroduction, but he threatened to block it using a state law that required the approval of the director of Fish and Game before any species could be reintroduced.[18]

Mealey's threat was an attempt to take advantage of the hostile political climate and give credence to the idea that the federal government was proceeding with reintroduction over the state's protestations. It exemplified the political resilience of the Old West, but for the moment, the threat was relatively empty. The USFWS would not release the final EIS for another few years, so his signature would not be required until after that point. His public declaration, however, was a signal of the political storm that was poised to consume reintroduction in the coming years.

At the beginning of 1998, Defenders of Wildlife and the National Wildlife Federation devoted most of their efforts to working with the U.S. Fish and Wildlife Service to mitigate the increasingly hostile political environment. In January, the USFWS made a plan to release the final EIS later that year and begin reintroduction by 1999. Defenders of Wildlife and the National Wildlife Federation still wanted the USFWS to release the final version as soon as possible, but the idea that a hastened schedule would avoid controversy was moot. Now bear advocates had to be much more strategic in order to court political support, or at least mitigate opposition from Idaho and Montana's volatile politicians. This thinking became especially pertinent as Senator Kempthorne emerged as the most likely candidate to become Idaho's next governor. Although Kempthorne had started taking jabs at reintroduction, the ROOTS coalition believed he privately endorsed it. He had not taken a firm public position, and bear advocates wanted to keep it that way. If the USFWS released the final EIS before the November election, the political firestorm that would incur would undoubtedly force Kempthorne to take the safer political route. They also believed that once he became governor, he would be in a position to be more supportive. With this in mind, Hank Fischer, Tom France, Chris Servheen, and Jamie Clark, the new director of the USFWS, decided to delay the release of the EIS until after the election, in December.[19]

While Kempthorne's temporary silence was fairly easy to engineer, bear advocates had to brace themselves to fight another round of riders. Fischer and France knew that Senators Burns and Craig planned on introducing another bill, similar to the one from the previous year, which would prevent the USFWS from spending money on reintroducing grizzlies to the Bitterroots in 1999. The previous year's bill had been largely symbolic because the USFWS had no intention of reintroducing grizzly bears in 1998, but they *had* hoped to begin reintroduction by 1999, so defeating this bill was paramount to the proposal's success.[20]

Over the course of 1998, Defenders of Wildlife and the National Wildlife Federation spent most of their efforts trying to defeat the rider, but as they were fighting that issue, they needed to convince the Fish and Wildlife Service to throw its full weight behind the project. Although the USFWS was the agency in charge of the environmental impact statement process, it was also vulnerable to the political pressures that were engulfing the project. So even though the Endangered Species Act and the Grizzly Bear Recovery Plan required it to recover grizzlies in the Bitterroots, the agency was susceptible to the rough political climate. Resistance from environmental groups that wanted the more biologically pure conservation biology alternative had already fostered some hesitancy within the USFWS, and as defiance from the conservatives escalated and confrontation beckoned, the USFWS proceeded timidly.[21]

Despite remonstrations from Defenders of Wildlife and the National Wildlife Federation, the USFWS never committed itself fully again for the remainder of the project, and even Defenders of Wildlife considered abandoning the project at one point. But with Fischer's insistence, the organization decided to see it through alongside the National Wildlife Federation, whose commitment never wavered. Their first avenue to fighting the rider was through Senator Baucus. Although the political climate had fostered cautiousness in him, Baucus still supported the project and he was the most likely member of Montana or Idaho's delegation to lead this cause. Baucus had been Montana's senator for twenty years and he was a member of the Environment and Public Works Committee. In this role he had previously demonstrated a willingness to fight for wildlife and endangered species protection. Although he was no friend of gray wolf reintroduction, in 1995 he had openly opposed Senator Jesse Helms's attempt to derail red wolf reintroduction in North Carolina. In a meeting with

Baucus that June, the ROOTS coalition reminded him of this battle and urged him to stand up again on behalf of wildlife. As a result, Baucus agreed to write a joint letter with John Chaffee of Rhode Island and Harry Reid of Nevada to the Senate Appropriations Committee urging them not to include the rider.[22]

Once the coalition confirmed Baucus's support, they pursued additional avenues to fight the rider. ROOTS members met with Idaho Representative Mike Crapo and Montana's freshman Republican congressman Rick Hill, insisting they oppose it. Crapo privately endorsed the citizen management alternative, and Hill had proven sympathetic to the idea of citizen management, but neither was willing to make their positions public. ROOTS members also met with Senator Burns in Washington in hopes of convincing him not to introduce another rider and let the EIS process unfold as intended. At the meeting, Burns said that his mind was not made up fully to oppose reintroduction and that more support from the Bitterroot Valley could change his mind. Baucus and Racicot also talked with Burns on separate occasions, attempting to discourage him from pursuing the rider, but Burns's position was more staunch than he let on. In late June, Burns and Craig again introduced a rider barring the USFWS from funding grizzly bear reintroduction. The White House put it on its list of environmental riders to oppose and threatened to veto the bill, but President Clinton eventually relented and signed the appropriations bill anyway. There would be no reintroduction in 1999.[23]

Despite this setback, the Fish and Wildlife Service still had seventy-five thousand dollars to complete the EIS and it continued to refine the final plan. The agency had fallen behind the schedule set after the release of the draft EIS, and the majority of the delay resulted from the controversial political climate. Earlier in 1998, Custer County, Idaho, which lay at the southern end of the experimental area, passed an ordinance barring grizzlies from central Idaho. At the outset of 1999, Montana followed suit when Ravalli County's representative Allan Walters submitted a bill to the state's legislature that prevented the bear's release, citing the idea that the Endangered Species Act was simply a tool to lock up land.[24] By this point, controversy was unavoidable. Neither of the resolutions was binding because the Endangered Species Act was a federal law, but they were proof that tension was increasing.

Amidst these obstacles came some good news for grizzly bears. Mark Boyce, a researcher from the University of Wisconsin–Stevens Point, who had been commissioned to complete the habitat study included in Burns and Craig's 1998 rider, published his findings. According to Boyce's habitat study, the Bitterroots could support more than 300 grizzly bears, which was more than the 280 bears that the draft EIS had estimated. Even though Burns and Craig had sponsored the study, they immediately attacked its findings. They questioned Boyce's methodological approach, saying his comparison between the Bitterroots and Yellowstone was misguided because the areas were ecologically disparate. Other people maintained that the region would not be able to support bears because the historic salmon runs no longer existed. In response, Chris Servheen reminded critics, "In the Rocky Mountains, there are no bears that eat salmon anymore, but we still have grizzly bears in the Rocky Mountains."[25]

The one point of Boyce's findings in which Burns, Craig, and other critics of reintroduction found comfort was his prediction that grizzly bears had a one in a million chance of going extinct over the next one hundred years. From this point, some Idaho and Montana politicians wanted Bitterroot reintroduction to be called off and for delisting of the grizzly bear throughout its range to begin. Bear advocates were quick to point out that the study actually concluded that the chances of extinction ranged from one in three hundred thirty to one in five million, and that it estimated that Bitterroot reintroduction would increase the species' chance of survival by a minimum of 88 percent. Others also noted that Boyce's findings assumed that ecological conditions would remain constant over that period, something that might not be the case if the bears were delisted.[26]

As with wolf reintroduction, political support remained the essential factor that would make or break reintroduction, and the options for grizzly advocates were increasingly limited. Baucus had written a letter the previous year to convince the Senate Appropriations Committee not to include the rider. However, his letter only expressed support for letting the EIS process unfold as scheduled. He did not come out and support reintroduction outright, and so to combat the added momentum that grizzlies' detractors had gained, bear advocates needed to find a more dedicated booster at the federal level. Secretary of the Interior Bruce Babbitt was their last hope. A native of Arizona, Babbitt had been a friend to the environment during his time in office. His support had helped make wolf reintroduction possible, and throughout his tenure, he made other reforms that had revised federal land management policies in the mold of the environmental movement. In 2000, he would establish the National Landscape Conservation System, a massive conglomeration of public lands in the West designated for protection and restoration. According to Hank Fischer's thinking, if Babbitt got behind the project, he could convince Jamie Clark, the director of the Fish and Wildlife Service, to proceed. She had voiced support for the plan at regular intervals, but continued to drag her feet. Part of this delay came from a desire to redirect the agency's focus. Critics had often complained that the USFWS and Endangered Species Act only protected charismatic megafauna and ignored less prominent species. However, the coalescence of Old West interests and pressure from Kempthorne added to Clark's unwillingness to proceed. The ROOTS group met with Babbitt in Washington, but he showed little inclination to push Clark on something she did not want to do.

In the midst of this unfortunate series of events, bear advocates briefly had a reason to celebrate. In 1999, Senators Burns and Craig introduced another rider that would prevent reintroduction in 2000, but President Clinton finally took a harder position and forced Republicans to drop this, as well as other antienvironmental riders.[27]

Even so, the good news did not last. The EIS still remained lost in Washington's bureaucratic channels, and Defenders of Wildlife and the National Wildlife Federation struggled to find ways to push the issue further. That fall, reintroduction's prospects took another hit. Newly elected Governor Kempthorne, who had taken office in January 1999, finally took a public stance on the issue. Despite Fischer's prediction that he would be more supportive of the plan once he became governor, Kempthorne went the other way, carrying the torch for his Old West constituents. In a September 1999 press release, he minced no words. "I want to make it perfectly clear that I oppose any reintroduction of grizzly bears into Idaho ... period," he said. While it surprised Fischer, Kempthorne's new position was consistent with Idaho's recent political history. [28]

Governor Cecil Andrus, who preceded Phil Batt, had exemplified the New West mindset in his time as secretary of the interior under President Jimmy Carter. During his time in Washington he helped create fifty-six million acres of wilderness in addition to twelve national parks in Alaska, but as soon as he returned to Idaho, Old West interests forced him to assume a much more conservative position on environmental issues. He supported salmon recovery, but only because he was an angler, and because protecting salmon fit comfortably within the parameters of conservation, it was politically safer.[29] In contrast, bears and wolves were not covered by the Old West's conservation ideology, and Andrus opposed their reintroductions vigorously. Kempthorne was no more immune to these forces, so when he returned to Idaho, the state's conservative political culture forced him to harden his position against the ROOTS plan.

Montana's political culture was similar in many ways, but with more urban centers and a larger New West population, it was more liberal and more representative of the region's changing demographics than was Idaho. In 1996, 33 percent of Idahoans voted for President Clinton while 41 percent of Montanans voted for the incumbent Democrat. Similarly, in 2000, both states voted for Republican George W. Bush, but only 58 percent of Montanans voted for Bush whereas 67 percent of Idahoans did. The states' elected officials, especially their governors, reflected these subtle distinctions.[30]

Racicot and Kempthorne shared many similarities. They were both Republican governors from the interior West, and after 2000, their respective relationships with George W. Bush catapulted them onto the national stage. But they did differ, and their contrasting qualities reflected the unique personas of their respective states. The more moderate Racicot truly believed in consensus and bipartisanship, whereas the more ideologically driven Kempthorne gave lip service to these political watchwords while consistently opposing meaningful environmental reforms; and such was the difference between the two states as well. Both Idaho and Montana maintained strong ties to their Old West identities, but in Idaho, the Old West had retained more influence than it did in Montana. For the Old West, there was almost no room for compromise because any relinquishment of power was viewed as a slippery slope that would lead to the unraveling of the region's social and economic order. The New West, on the other hand, understood its status as the up-and-coming power and was willing to make concessions to achieve smaller victories. The larger faction of New Westerners in Montana made more likely the election of politicians who genuinely embraced compromise and consensus as useful political tools instead of employing them as self-serving maxims while adhering to rigid dogma.

Reflecting their inflexible adherence to the Old West's ideology, Idaho's other politicians joined Kempthorne in 1999, altering their positions or ramping up their campaign against reintroduction. Rep. Michael Crapo, who had been one of the project's most adamant supporters a few years earlier, had all but turned against it, and in rural Idaho, Helen Chenoweth and Larry Craig helped citizens form an antigrizzly coalition. At a meeting in Salmon, the group passed a resolution stating that grizzly bears had never resided in the Bitterroots and if the habitat actually could support bears, they would be there already. At the meeting, in a brief moment of absolute clarity, one of Senator Craig's staffers quipped, "This is more than just an experimental introduction, it is a bid for controlling the entire western way of life."[31] The power of this insight was probably greater than its speaker could have surmised as it provided a context for every other complaint they made. Indeed, reintroduction was about control of the region, and Old Westerners knew that even if not all their fears came true, reintroducing grizzly bears would be another shining example of how their influence writ large had dwindled. A little more than four years had passed from the time the ROOTS coalition convinced Congress to fund the EIS even as last-minute lawsuits were trying to stop wolf reintroduction. Consensus had made that possible, but times had changed significantly, and no degree of compromise could mask the larger implications of a grizzly bear reintroduction.

To complicate matters further, hard-line environmentalists and supporters of the conservation biology alternative began attacking the legality of the experimental designation with renewed vigor. A federal court ruled in 1998 that the experimental designation of wolves was illegal because fully protected wolves could potentially migrate into the experimental recovery area, in which case their protected status would be downgraded. Serveen insisted that this would not affect Bitterroot grizzlies because no grizzlies resided in the Bitterroots, and of the 550 bears the USFWS had radio-collared since 1975, not a single one had migrated between recovery areas. Even so, some environmentalists had their doubts. Since the release of the draft EIS, Mike Bader and other environmentalists had occasionally suggested that a small, remnant population of grizzlies still persisted in the Bitterroots, but the USFWS had not substantiated any sightings since Bud Moore saw the last tracks in 1946.

The recent court ruling on wolves breathed new life into this campaign, so in the fall of 1999, the Alliance for the Wild Rockies, along with the Sierra Club, Friends of the Bitterroot, Great Bear Foundation, and the Craighead Wildlife-Wilderness Institute, announced the launch of the "Great Grizzly Search." The search consisted of sending volunteers into the Bitterroots to collect hair and scat samples and arming them with pocket guides to help distinguish grizzlies from black bears. The search also included aerial reconnaissance scanning for bears and their dens.[32] According to this strategy, once evidence of grizzly bears surfaced, one or more groups could sue the USFWS to stop the experimental reintroduction.

Like every development in this slowly unfolding saga, the launch of the Great Grizzly Search brought controversy. Chris Servheen was immovable on this issue. "There's no evidence of grizzly bears in the Bitterroots. None. Zip. Nada," he commented, before adding, "We have no reason on God's green earth to hide evidence of grizzly bears. What purpose would I have to hide evidence of grizzly bears? Grizzly bears are what I do." Nevertheless, some environmentalists were convinced of the bear's presence in the Bitterroots. Pointing to the weighty stack of reported grizzly sightings from the 1960s onwards, Larry Campbell of Friends of the Bitterroot commented, "What does it take for agencies to admit it's been documented?" Even with these reports, Servheen refused to budge from his position unless some hard proof surfaced, whether it be a photograph or cast of a print. Distinguishing between black bears and grizzlies could be difficult, and he, as well as other bear biologists, did not trust these reports as they had spent many hours chasing false leads. "I've followed up on everything, and there just hasn't been any solid evidence," claimed bear biologist Dan Davis, who had conducted one of the preliminary habitat studies in the Bitterroots back in 1991 and had spent the years since continuing to research grizzly bears.[33]

In the 1980s, a hunter had seen a bear feeding on a carcass in the Bitterroots and had reported the sighting to Davis. An observer returned to the scene and watched the carcass through a spotting scope for three weeks, but only saw black bears return to it. Another time, a friend of Davis's ran out of the Bitterroots to tell him about a massive paw print he found in the snow that he thought could be a grizzly. But by the time Davis and his friend reached the track again, it had melted and was useless. The same year the search began, a group of hunters were certain they saw a grizzly near Salmon, Idaho, and state and federal biologists collected hair samples from a nearby tree rub. They sent the samples to a lab for testing, but DNA analysis determined they came from a black bear. According to Steve Nadeau, who was a bear biologist for Idaho Fish and Game and part of the recovery team, "These things start to wear on you after a while." Even so, people who had claimed to see grizzlies insisted on the veracity of their sightings and refused to let government officials question their observations just because they did not hold PhDs.[34] For them, the Great Grizzly Search was a chance to substantiate what they already knew to be true.

Because the organizations running the search had a distinct agenda and had a lot invested in producing positive results, Hank Fischer was not too happy about its prospects. He was convinced it would turn up something "even if they manufacture it," and he warned his ROOTS colleagues to be prepared. Despite this concern, the search continued for three consecutive summers between 1999 and 2001 and found nothing. Volunteers collected numerous hair samples that they sent to University of Idaho labs to be tested, but every test came back "black bear."[35] The search never realized its ultimate goal of stopping the reintroduction of an experimental population through legal channels, but in the meantime, it put everyone involved on edge. The dozens of enthusiastic CBA supporters roaming through the Bitterroots was great publicity and reinvigorated opposition from the left, which had fallen relatively silent over the previous year.

Finally, in March of 2000, the USFWS released the long awaited final EIS. The citizen management plan remained the preferred alternative, and the entire statement was none too different from the draft version, but it did incorporate many of the suggestions made by Governor Racicot and the ROOTS coalition. Although the final version shared many similarities with the draft, it added two nonvoting scientific advisers to the citizen management committee, mandated that grizzlies found outside the recovery area would only be removed if they threatened people or livestock, and included a clause that required the citizen committee to adhere to the best available science. The final EIS analyzed two additional alternatives as well. Alternative 1A, an amended version of the citizen management alternative, would have reintroduced a nonessential, experimental population but would have let the Fish and Wildlife Service control management. Alternative 4A was a scaled-down version of the conservation biology approach that did not include the far-reaching habitat restoration initiatives. The USFWS rejected both alternatives because it felt that citizen management was necessary to achieve success.[36]

Naturally, the release of the final EIS set off another flurry of comments and opinions on the prospect of reintroduction. While much of the previous resistance to grizzly bears had hinted at the resentment created by the divide between the Old West and the New West, commentators began directly invoking this split. "What we don't need is the Forest Service, U.S. Fish and Wildlife Service or the Environmental fringe dumping a massive problem on the citizens of Idaho," commented one observer. Another added, in what could only be a reference to New Western ideology, "The U.S. Fish and Wildlife Service is more sympathetic to dreamy, New Age dialectics founded upon willful ignorance and pseudospirituality than scientific calculations." One editorialist added, "The yuppies clamoring for a wilderness experience in bear country can still grab their bear bells and hike a trail leading away from the safety of Glacier's Going- to-the-Sun Highway." In one final example, explicitly invoking the split between the New West and Old West economies, one Old West ideologue offered, "If you tried to organize meetings with your typical left winger, you would find that they have way more time than myself. The people that are really affected are so busy with their heads down working all the time, and we don't have the luxury of organizing and getting vocal."[37] Few of these comments were substantive or addressed the merits of the citizen management alternatives. Rather, they all focused on broader issues related to the dynamic between the Old West and the New West and demonstrated how difficult it was to have an honest, straightforward conversation about the issue at hand when the groups involved were so diametrically opposed to one another.

From the left, bear advocates were just as willing to make appeals to this divide. Referring to the primary alternatives, Mike Bader charged that the two proposals "have a different sense of where we are in history." Bader believed that the citizen management approach had made far too many concessions in an effort to appease the Old West, and he thought that the time had come when the New West was powerful enough to push through a biologically stronger plan. Because some in the New West were very much aware of this divide and the types of criticisms that Old Westerners made of them, they employed reverse psychology. Montana and Idaho had seen an influx of new residents coming from California, and there was nothing Old Montanans disdained more than a sports car carrying a couple of cappuccino-drinking Californians. "I'd just as soon not become like California in one more way," remarked Wild Forever's Brian Peck, referring to the fact that California had killed its last grizzly almost eighty years earlier. Similarly, in 1997, Hank Fischer off-handedly dismissed Helen Chenoweth's antibear rhetoric by saying that she "sounds like a Californian." Another observer taunted supporters of the Old West by making an appeal to their sense of rugged individualism, "Let the West be the West again. Be brave and take a chance." Also reflecting the fact that the split over bears fell along Old West–New West lines, and that New Westerners tended to be younger, one Idaho poll, which had support for reintroduction in Idaho at 43 percent, found that number had jumped to 58 percent among Idahoans ages eighteen through thirty-four.[38]

Rhetoric aside, the law required the Fish and Wildlife Service to recover grizzly bears in the Bitterroot ecosystem, and because most biologists agreed that recovery would not happen without reintroduction, the plan had both science and the law on its side. However, it still needed funding, and over the past two years, the bears' opponents in Congress had been able to delay the project by limiting funding, just as opponents of wolves had done for nearly a decade. With this in mind, Fischer, France, and other advocates of the citizen management alternative wasted no time after the final EIS's release before beginning their last campaign to satisfy Marc Racicot's lin-

gering concerns and secure funding. In April the outlook for funding looked grim as the USFWS cut the entire grizzly bear budget by
$126,000. By May, however, Fischer felt encouraged. He had met
with Jamie Clark, who, for the first time, made a strong indication
of support. Still, she conceded that funding would not be available
until after 2001, and they all agreed not to move forward with reintroduction until funding was assured. For the previous few months,
Defenders of Wildlife and the National Wildlife Federation had discussed funding the reintroduction themselves, but at the meeting
with Clark, they agreed that private funding should be limited to the
education and sanitation programs.[39]

The other strategy that came out of the meeting with Clark was to
establish the citizen management committee as quickly as possible
and compel Racicot and Kempthorne to make their appointments.
Once the committee was operational, people who doubted its efficacy
would have a chance to see it in action in a low-stakes environment
before grizzly bears were on the ground. Furthermore, the theory
reasoned that a functioning committee would make funding easier
to secure because a major element of the project would already be in
place. Fischer met with Crapo, Burns, and Craig to discuss this possibility, and surprisingly, they all showed some level of enthusiasm
for this approach. Similar to the closing years before wolf reintroduction, these politicians started to accept that reintroduction would
happen and realized they would be wise to get on board. Within a
few months, Burns included a one-hundred-thousand-dollar appropriation in the federal budget to be applied toward reviewing Boyce's
habitat study. The review was intended to be less of a serious review
than an easy opportunity for the citizen management committee to
get off the ground and prove that it could function as intended.[40]

The citizen committee had been the reason Racicot had supported
the plan from the beginning, and even before the release of the final
EIS, he had demonstrated a willingness to make appointments to
it. However, once Kempthorne publicly opposed reintroduction,
the Idaho governor never wavered. "Whenever there's an encoun-

ter between a human and a grizzly, the human does not fare well," Kempthorne claimed. He later declared, "I oppose bringing these massive flesh-eating carnivores into Idaho." One commentator shot back, "The people need protection from the agonizing words of Dirk Kempthorne"; but the governor could not be appeased or quieted. When the USFWS finally published the Record of Decision and Final Rule, on November 17, 2000, choosing to implement the citizen management alternative, Kempthorne immediately threatened to sue the Fish and Wildlife Service to stop the reintroduction.[41]

The USFWS sent Kempthorne a letter urging him to temper his resistance, embrace the citizen committee, and appoint members to it, but that same day, Idaho's Constitutional Defense Fund announced its intention to hire an outside legal team to fight the reintroduction. Then, the day before Republican president George W. Bush took office, Idaho filed a lawsuit alleging that reintroducing grizzly bears would threaten the safety of residents and visitors and the decision infringed on the state's sovereignty. The lawsuit was much more of a political statement than a legal one. All the precedents were against it. Not only was the decision solidly a federal one, but the USFWS had held numerous meetings in Idaho and had received more than twenty-six thousand comments from the public. Additionally, the citizen management plan was created to undercut just such a complaint.[42]

The lawsuit's merits quickly proved moot. The action, along with Kempthorne's rabid rhetoric, was enough to get the attention of President Bush and newly appointed Secretary of the Interior Gale Norton. Norton insisted that she was "committed to working on recovering and growing the number of wild grizzlies in the forty-eight states," and Bush himself touted the merits of finding local solutions to natural resource problems. Nevertheless, Tom France got word that Kempthorne had the administration's ear, and most bear advocates had already accepted that Bush's election meant that grizzly reintroduction was doomed.[43]

In June 2001, Norton announced her decision to withdraw the Final Rule, citing the lack of local support for the plan. Bear advocates were outraged. The decision, they protested, was based on politics, not science, and therefore illegal under the Endangered Species Act. France called the decision "short-sighted and stupid" and concluded that it not only hurt grizzlies but "the whole effort of finding compromises and common-sense solutions to endangered species management." Sterling Miller dubbed it "a travesty of law, science, and the public process."[44]

A sixty-day public comment period followed Norton's decision, giving the public another formal opportunity to weigh in. Realizing the precedent this set, even the Sierra Club and other groups that had fought for the conservation biology alternative wrote letters disparaging Norton's decision, and when the USFWS tallied the results in October, the decision was nearly unanimous. Of the twenty-eight thousand people who submitted comments, 98 percent wanted reintroduction to proceed. Of Idaho and Montana commentators, the percentages were 98 percent and 93 respectively. Although these numbers did not truly represent public opinion at large, any chance that Norton could reverse the USFWS's decision because of a lack of public support evaporated. However, money remained an uncertain prospect. The project had struggled to secure funding even while the relatively bear-friendly Clinton administration was in office, and the Bush administration clearly had no plans to stick its neck out to secure money. "Obviously," said Tom France, "there are new spending priorities, at least for the short term. But we can wait a year or two or three to move forward with this program. We can wait. That is far different than permanently abandoning a program." Despite this optimism, Fischer, Servheen, and other bear advocates saw the writing on the wall.[45] Griz football could not be beaten, but Bitterroot grizzlies had lost. If the bears were going to return to the Bitterroots, they would have to do it on their own.

The USFWS had released the final decision shortly after the election, while George Bush and Al Gore still battled over contested votes in Florida. Part of the reason for the timing of the release was to protect the decision from appearing as a reaction to the election's results.

However, even before the disputed election, many people involved with the project knew reintroduction's implementation would hinge on its results.[46] Reintroducing grizzlies had never been purely a biological issue. Unbeknownst to any actual grizzly bears, their presence in one of their native ranges had become a highly controversial, partisan issue that split along the growing divide between the Old West and New West. For a number of years, the ROOTS coalition had been able to forge a compromise that temporarily allowed the issue to transcend this divide, but in the end, the ideological distance between the two groups proved too great for the compromise to bridge.

Despite their personal beliefs and, in certain instances, common sense, western politicians increasingly felt pressure-bound by tradition. To forsake the people and ideology that had shaped the region over the previous hundred years was untenable. Grizzly bears may not have been as dangerous as they originally believed, the citizen management plan may have alleviated feelings of federal overreach, and the experimental designation may have ensured that extractive industries would not be locked out of their forests, but those issues were beside the point. Reintroducing grizzly bears was about who controlled the West, and while the success of wolf reintroduction may have signaled a temporary change, the failure of grizzly reintroduction signaled that the Old West's hold over the region was still strong.

Conclusion

By 2007 Bitterroot grizzly reintroduction was all but forgotten. Montana, Wyoming, and Idaho were working toward delisting Yellowstone's grizzly population, and the attention of bear advocates had shifted toward fighting to retain that population's protected status. So when a black-bear hunter from Tennessee traveled into the Great Burn region of the northern Bitterroots on a guided trip that September, grizzlies were the last thing on his mind. The hunter and his guide were in Idaho, within the boundaries of the proposed experimental recovery area, roughly twenty miles north of the Selway-Bitterroot Wilderness. The outfitter had set up a bait station to attract nearby bears, and as the hunter quietly waited over the site, a dark, 450-pound bear wandered into view. The guide had stepped away for a moment, and the out-of-state hunter had no reason to believe the large creature was not a black bear. Once the bear was in his sights, he pulled the trigger, killing the bear. The hunter could not have been more excited, but when he and his guide went to admire his kill, they quickly determined, to their surprise, that he had shot a grizzly.[1] To the shock of nearly everyone involved with Bitterroot grizzly recovery, the hunter was responsible for the first confirmed sighting of a grizzly in the Bitterroots since 1946.

News of the bear's demise traveled quickly throughout western Montana. While many people, including the hunter, lamented the death of a threatened animal, bear advocates were more excited about what this meant for the future of grizzlies in the Bitterroots. Despite his prior doubts and protestations that bears could not repopulate the region on their own, Chris Servheen understood the potential implications of this discovery. Using the DNA database they had developed over the previous decade of fieldwork, Servheen and the Fish and Wildlife Service discovered, to their amazement, that the bear had migrated 140 aerial miles from the Selkirk ecosystem. The bear's lengthy journey from the northern tip of Idaho to the Bitterroots further heartened grizzly supporters, and the USFWS quickly arranged for an official survey to take place in the Bitterroots the following summer. Unfortunately, the study, which included fifty-one motion-sensor cameras and hair-snag stations, produced no further evidence of grizzlies in the region.[2] Just as suddenly as bear advocates had a reason to hope again, they were forced back to square one.

Chris Servheen had hoped the incident would renew talks of reintroduction, but any such discussions faded into obscurity once again, and other wildlife issues took precedence in the public eye. In 2007 the USFWS delisted Yellowstone's grizzlies, but two years later, a federal judge put them back on the endangered species list. More significantly, wolves were delisted across the Northern Rockies after a tumultuous multiyear-long battle. Wolf recovery had progressed quicker than any supporters had hoped, and by the early 2000s, wolves had reached the demographic and population requirements established by the environmental impact statement for delisting to begin. Even so, environmentalists fought the action, arguing that the population's long-term stability was still uncertain. The federal government successfully delisted wolves in 2009, but a federal judge put them back on the list the following year. In 2011, however, Montana, Idaho, and Wyoming finally brokered the wolves' delisting, and the states assumed full authority over their management. After a court decision in 2014, Wyoming's wolves were relisted because the state's management plan was deemed inadequate. This was just one more step in the long, bitter fight over delisting that further polarized wildlife politics in the region and minimized any chances that the Bitter-root grizzly issue could regain momentum. For Old Westerners who already distrusted the federal government and resisted environmental reforms, the long, drawn-out process confirmed their worst fears of what the reintroduction of an apex predator would mean. The federal government's credibility in the region was already limited, and this lengthy process undermined its standing among former Sagebrush Rebels and hardened Old Westerners.

The implications of this fight were not lost on Chris Servheen who had still held out hope that Bitterroot reintroduction might occur. According to him, the protracted nature of wolf delisting "poisoned the well" for Bitterroot grizzlies, and he is convinced it foiled any possibility of reintroduction occurring in the near future. Nevertheless, Secretary of the Interior Gale Norton never withdrew the Fish and Wildlife Service's Record of Decision, so the plan remains official government policy that still awaits funding. Because more than a decade has passed, an additional habitat study would probably be necessary before the USFWS could move forward with reintroduction.[3] Even so, the prospect of having the states manage a new population of grizzlies is not nearly as palatable to environmentalists as it once was. Few in the environmental community have been satisfied with how Idaho, Wyoming, and Montana, to a lesser degree, have managed their wolf populations, and few would be eager to hand over the responsibility of a new grizzly population to these states.

In late 2014, the USFWS renewed talks concerning Bitterroot grizzlies, but the process is in the earliest stages and both the agency and advocates are moving slowly and methodically—aware of how delicate the issue is. In fact, more progress has been made in the North Cascades where discussions around augmentation also picked up again in 2014.

What was once going to be the model for endangered species restoration across the country and a way to return some civility to the West's political debates has been reduced to a footnote in a long list of things that could have been. While grizzly bears and a more sensitive relationship with the natural world were the most obvious losers, the failure of the ROOTS plan has had implications that stretch beyond the limited geographic and ecological boundaries of this story. In the West and among natural resource managers, the collaborative and consensus-based program could have helped diffuse the stalemate between the Old West and the New West, environmentalists and conservationists, urban and rural residents, and the many other groups that keep each other at arm's length. Instead, controversy continues to plague regional politics, and environmental issues remain susceptible to being undermined by the inflexible and dogmatic nature of ideological debate.

The mythology of the Old West still holds an iron-fisted grip over the region of which politicians and the public refuse to let go. With a few exceptions, like the Bakken oil boom in eastern Montana and western North Dakota, extractive industries have continued to decline, but the political support they receive remains vastly disproportionate compared to their share of the economy. Old Westerners have largely accepted the presence of new economic sectors, but they still do not want them to eclipse the region's original economy out of a belief that the region's exceptionalism is rooted in its extractive industries. In this estimation, they were partially correct. The West's vast natural resources—including its immense supplies of timber; unparalleled open spaces; harsh, unpredictable climates; and breathtaking mountains that stretch beyond the horizon—have always been a defining element in the region. However, while extracting and exploiting those resources may have been central to the region's character and originally made it what it was, this is no longer the case. Today, the resources themselves—the mountains, deserts, forests, and rivers—define the region, and protecting, preserving, and regenerating them has become the best way to preserve the West's exceptionalism. A study by a Bozeman-based economic research firm found that the prosperity of western towns will increasingly rely on their proximity to protected public lands. Similarly, a 2013 survey found that 70 percent of small business owners in Montana agreed that the state's outdoor recreation opportunities are a factor in locating or expanding there.[4] Extractive industries are prone to booms and busts, but the degree to which people from across the world value open spaces and wild places that are unique to the American West will be much less likely to fade and thus provide a much more sustainable source of income.

Even so, many in the Old West, fraught with anxiety over what these changes might mean, refuse to give ground. Conservatives, loyal to tradition and hostile to change, continue to wield power disproportionate to their demographic representation and ensure that actions such as grizzly reintroduction will not proceed without a long, bitter battle.[5] The New West has continued to gain ground with prospering businesses like Yoga on the Fly, a fly-fishing/yoga retreat designed specifically for women, but the region's politics lag behind, and the West remains a cultural, political, and social paradox.

In this way, the Bitterroot grizzly episode verified that the Old West's resistance to endangered species recovery went beyond economics. In contrast to earlier endangered species recovery programs whose opposition was cloaked in strictly economic arguments, such as the snail darter, spotted owl, and gray wolf, the plan to restore grizzlies enjoyed cooperation from the region's most dominant extractive industry. Instead, it was the real fear of a broader, more fundamental societal change that killed the project.[6] The Old West's success in defeating the proposal proved that even if its share of the region's economy had decreased, its political clout remained firm. Although many supporters of grizzlies wanted to believe the debate concerning reintroduction had shifted from *whether* to *how*, environmentalism's influence was not that universal in the Rocky Mountain West.

Just as much as this story offers a window into the ongoing struggle between the Old West and the New West for control of the region, it also suggests a lot about the efficacy of compromise as a tool in American politics. The ROOTS coalition had the potential to set a strong example of how diverse groups could work together, and its members hoped citizen management would be the future of endangered species programs across the country. It had created a new way to tackle controversial natural resource issues. Even its proponents were uncertain whether their model would work, but they knew it was worth trying given the intractable state of politics in the region at the dawn of the new millennium.

While our political system often champions compromise and pragmatism and views bipartisan efforts as admirable, this episode demonstrated that, in reality, those values take a backseat to codified ideology. Years of work, carefully crafted language, widespread public support, and thousands of dollars are simply no match for the power of a few determined individuals working from an intractable, unbending set of beliefs. The Sagebrush Rebellion and its later iterations complained of an overbearing federal government; and with this in mind, the ROOTS coalition created the idea of a citizen management committee so that locals would have a greater voice in the bears' management. Western politicians initially embraced this idea because it was exactly what they had been requesting, but reintroducing grizzly bears, under any circumstances, challenged their rigid principles, and in the end, they turned against it. Certain environmentalists were just as incapable of compromise. Because the ROOTS plan did not follow the strictest tenets of environmental thought, they refused to recognize its merits. While not as directly detrimental to the project's success, their refusal to compromise was nearly as damaging and demonstrated that from both sides, consensus and collaboration were trendy theories that sounded good in speeches, but ultimately paled in comparison to dogmatic ideology.

Since the collapse of the ROOTS plan, however, the environmental movement as a whole has moderated once again. Groups like Alliance for the Wild Rockies have faded, while organizations like the Sierra Club and Greater Yellowstone Coalition have moved more to the center. Increasingly, New Westerners have realized the need for compromise, and consensus and collaboration have become more prevalent tools in natural resource management. Compromise and consensus emerged as possible solutions to complex issues in the early 1990s, but as this story demonstrates, change happens slowly. When they developed the ROOTS plan, Hank Fischer and Tom France were at the forefront of natural resource policy, but since then, these strategies have become much more mainstream. In Montana's rural Blackfoot Valley, northeast of Missoula, a coalition of landowners, state, and federal agencies have united under the Blackfoot Challenge and have brought many environmental reforms and ecological restoration efforts to the valley. In Indiana's Charles C. Deam Wilderness, collaboration between hikers and horseback riders facilitated a compromise between the two groups; while in the Arctic, cooperation between hunters, biologists, and indigenous populations helped forge a solution concerning caribou management that met each group's needs.[7]

While many people—including Mike Bader, who has reaffirmed he is pleased that reintroduction failed—continue to find fault with this approach because of the biological concessions it requires, it has undoubtedly eased controversy and helped foster solutions where none had previously existed and has become a popular tool of many land managers. Strangely enough, the idea of compromise has embedded itself so deeply in the minds of some advocates that it has ceased to be a short-term necessity and has become standard operating procedure. While it is indeed an important tool, environmentalists must remain diligent and continually pressing the boundaries of what is possible. Even so, no one has again proposed citizen management for endangered species management, even though citizen participation has become a more prominent tool in other areas. However, as wolf delisting demonstrated, ideology is still capable of consuming and sidetracking any issue, and some matters will never be able to avoid the vitriolic debate that prevents productive and substantive conversations.

No one issue doomed Bitterroot grizzly recovery. Rather, a number of historical events—the listing of the spotted owl, the success of the highly controversial wolf reintroduction, and the rise of the New

West—set the stage for the architects of the ROOTS plan to develop the proposal they did. However, there were a number of other powerful forces that also led to the plan's demise—pushback by western land owners against endangered species recovery, a nationwide conservative shift, bureaucratic delays, lack of trust in the federal government, and local resistance to broad-based cultural changes that were challenging the West's ruling order. All these events were unique to that moment in history, and in this way, few other episodes provide a better lens for understanding the American West at the turn of the twentieth century and its cultural, social, and political struggles.

Environmental historian Dan Flores views the original eradication of grizzly bears in the West over the nineteenth and early twentieth centuries as a desire on the part of human beings to kill the wildness within; and keeping grizzlies out of the region undoubtedly represented an urge on the part of the Old West to maintain ideological ties with this past. By prioritizing society's progress and denying their own place in the natural world, Americans, especially westerners, justified this indiscriminate destruction.[8] Economics drove Gifford Pinchot's brand of conservation, thus leaving the primal instincts that dictated the destruction of predators intact. In some ways, it even encouraged and codified this behavior. Environmentalism challenged this mindset as it called for the protection of these species because of the ecological and spiritual roles they fulfilled. Still, backlash to the environmental movement in the form of the Sagebrush Rebellion and Property Rights Movements proved that these values were not fully ingrained in the American mind. By the time grizzly reintroduction in the Bitterroots came to a screeching halt in 2001, more than thirty-five years had gone by since Congress had passed the Endangered Species Act, and twelve years had elapsed since environmental historian Roderick Nash declared the law a moral victory for wildlife and a signal of the country's recognition of natural rights for animals. But the act was just a law, and its power paled in comparison to the entrenched societal values that placed humans above all other creatures.[9]

Fewer than a hundred years ago, humans hunted grizzlies to extinction in the Bitterroots and across the large majority of their range. Evolving environmental ethics challenged the philosophy that inspired those actions, but the persistence of nineteenth-century theories regarding humans' relationship with the natural world have kept the Bitterroots devoid of grizzlies even though the habitat remains ideal. As a result, the bear remains on the outside of one of its historic homes and one of the last places where it could thrive. Minette Johnson regrets this immensely. Remembering how close the plan came to being realized, she lamented, "Maybe there would be bears back now. Maybe my kids could go hiking in the Bitterroot and see grizzly bears, which would be an amazing experience. But it's not going to happen." For anyone who spent as much time as Johnson and other environmentalists did advocating for reintroduction and the richer relationship with the natural world that it would have fostered, remorse remains as their only solace.

The year after I stood atop Cha-paa-qn and gazed longingly south toward the Bitterroots, wondering why a region so extensive and so wild lacked the very animal that exemplifies wildness above any other, I made my first sojourn into its forested valleys and up to its many crystal-clear lakes. As I walked, I secretly hoped that I would stumble upon a grizzly and reinvigorate the debate once again. I, of course, knew how unlikely this prospect was, and after a number of days and nights spent wandering through the mountains that summer, I begrudgingly accepted that my search was in vain. I wanted to the feel the rush of knowing that at any moment I could round a bend and come face-to-face with something more powerful than me. I tried to be acutely aware of my surroundings, so that I would not be surprised when that moment came, and I yearned to feel the close connection with the landscape that such a level of awareness brings. Instead, I ambled, carefree past ancient red cedars and granite walls that have stood for millennia, and while I cherished every step and every breath, it felt empty. I could sense that something was missing.

Notes

Abbreviations

BS   Boise State University Archives and Special Collections
CAC Cecil Andrus Collection
HFP Hank Fischer Papers
ML   Archives and Special Collections, Maureen and Mike
     Mansfield Library, University of Montana–Missoula

Introduction

1. Information about settlers killing off bears is from Moore, *Lochsa Story*, 267–69, 278. The effect that roads have in keeping grizzly bears from migrating is from Servheen, *Grizzly Bear Recovery Plan*, 23–24.
2. Fischer, "Moving Past the Polarization," 1–4.
3. For the plan, see U.S. Fish and Wildlife Service, *Final Environmental Impact Statement*, xvi–xxiv. For complaints from conservatives, see Cawley, *Federal Land, Western Anger*, 76.
4. Fischer and Roy, "New Approaches to Citizen Participation," 603–6.
5. Fischer, "Moving Past the Polarization," 10.
6. Bader, Garrity, and Bechtold, *Conservation Biology Alternative for Grizzly Bear Population Restoration*.
7. For examples of what the Old West is, see Wiltsie and Wyckoff, "Reinventing Red Lodge," 130–36; Amundson, "Yellowcake to Single Track," 150; and Culver, "From 'Last of the Old West,'" 164. For

information about conservation, see Pinchot, "Ends and Means," 58–64.

8. Hays, *Beauty, Health, and Permanence*, 115–16; and Rothman, *Devil's Bargains*, 24. For the rise of the new upper class in the West, see Brooks, *Bobos in Paradise*, 25–28; and Travis, *New Geographies of the American West*, 22–26. For more about the rise of the environmental movement, see Rothman, *Greening of a Nation?*, 2–3

9. For conservationists' reaction to environmentalism, see Hays, *Beauty, Health, and Permanence*, 65–67; Rothman, *Greening of a Nation?*, 5. Information about the Sagebrush Rebellion is from Cawley, *Federal Land, Western Anger*, 1–2, 9, 71–72, 166–67.

10. Flores, foreword to Nicholas, Bapis, and Harvey, *Imagining the Big Open*, viii; Nicholas, "1-800-SUNDANCE," 262, 65.

11. U.S. Fish and Wildlife Service, *Final Rule on Establishment*. For the Bush administration's actions, see Associated Press, "Norton Proposes Scrapping Griz Plan," *Missoulian*, June 21, 2001. For observers' reactions, see Hank Fischer, interview with the author, March 2, 2012, in possession of the author. Importance of the election is from Dan Hansen, "Kempthorne to Grizzlies: Keep Out!" *Spokane Spokesman-Review*, November 17, 2000. Kempthorne's quote is from Sherry Devlin, "Grizzlies OK'd for Bitterroot," *Missoulian*, November 17, 2000.

12. Lapinski, *Grizzlies and Grizzled Old Men*, 6–13.

13. Mighetto, *Wild Animals and American Environmental Ethics*, 91; Biel, *Do (Not) Feed the Bears*, 21–23, 26, 41.

14. The implication of reintroduction is from Stout, "Reintroduction of Grizzlies," 28; and Jones, "Way Out West," 28. The first quote is from "Grizzly Bear Population Recovery." The second quote is from D. F. Oliveria, "Bear Compromise Shows There's Hope," *Spokane Spokesman-Review*, July 26, 1995.

15. Ted Kerasorte, "Wolves Bring Yellowstone to Vivid Life," *High Country News*, June 26, 1995; Askins, "Releasing Wolves from Symbolism," 15; Michael Milstein, "The Wolves Are Back, Big Time," *High Country News*, February 6, 1995.

16. For the West's economy, see Travis, *New Geographies of the American West*, 22–26. For Old West anti-authority stance, see Brooks, *Bobos in Paradise*, 223. For ideas of authenticity, see Flores, foreword to *Imagining the Big Open*, viii.

17. Patterson, *Restless Giant*, 324.
18. For articles from out of state, see "Bear Essentials: Blurring Battle Lines," *Arizona Republic*, May 25, 1997; Ken Olsen, "Survey: Bring Grizzly Bears Back to Idaho Wilderness," *Anchorage Daily News*, June 22, 1997; "Poll Shows Support for Grizzly Reintroduction," *Roseburg (OR) News-Review*, June 22, 1997; Jim Robbins, "Plan to Repopulate Grizzlies Gains Support," *New York Times*, April 27, 1997; and Tom Kenworthy, "Politics Imperils Uncommon Alliance's Plan to Find Grizzlies a Home," *Washington Post*, October 12, 1997. See also Knibb, *Grizzly Wars*.
19. Jones, "Way Out West," 38–40; Wilkinson, "Paradise Revised," 44.

1. Grizzly Americana

1. Wright, *Grizzly Bear*, 63–68.
2. The biographical information about Wright can be found in Wright, *Grizzly Bear*, vi, vii, 3–10. For the stories of his grizzly bear hunts, see Wright, *Grizzly Bear*, 56–61, 91, 95.
3. Information in the first part of the paragraph comes from Schullery, *Lewis and Clark Among the Grizzlies*, 4. For the second part of the paragraph, including many accounts of the expedition shooting bears, see Flores's chapter, "Dreams and Beasts," in *Natural West*, 77–80. For the quote, see Flores, "Dreams and Beasts," in *Natural West*, 78.
4. Schullery, *Lewis and Clark Among the Grizzlies*, 128–129. The fact about the bears killed around Kamiah comes from Melquist, *Preliminary Survey*, 14, HFP, ML.
5. Schullery, *Lewis and Clark Among the Grizzlies*, 129; Flores, "Dreams and Beasts," in *Natural West*, 86–88; Knibb, *Grizzly Wars*, 18. Also see Miller, "A Grizzly's Sly Little Joke," 331–32, and Lummis, "Begging the Bear's Pardon," 333–34. For the Nez Perce tale, see Clark, *Indian Legends of the Northern Rockies*, 44–45.
6. For the naming of the bear by Ord, see Murray, *Great Bear*, 235.
7. U.S. Fish and Wildlife Service, *Final Environmental Impact Statement: Grizzly Bear*, 6-25–6-28.
8. Washington Irving's story is from Irving, "Adventures of William Cannon and John Day with Grizzly Bears," 27–28.
9. Ruxton, "The Saga of Hugh Glass," 51–54.

10. For details concerning nineteenth century attitudes toward bears, see Lapinski, *Grizzlies and Grizzled Old Men*, 6–13; and Robinson, *Predatory Bureaucracy*, 12–13.

11. The first statistics for the Hudson's Bay Company and George Yount come from Knibb, *Grizzly Wars*, 20. The last half of the paragraph comes from Robinson, *Predatory Bureaucracy*, 31, 118.

12. For information about Grizzly Adams, see Murray, *Great Bear*, 236. Mills, "Trailing Without a Gun," 182–92. For the quote, see Mills, *Grizzly*, 284.

13. Roosevelt, "Hunting in the West," 125. For details concerning the state-by-state dates when grizzlies went extinct, see Flores, *Natural West*, 85; and Robinson, *Predatory Bureaucracy*, 285, 298. The information concerning the hunt in Arizona can be found in Lapinski, *Grizzlies and Grizzled Old Men*, 18. For the details about their range and current population numbers, see Peacock and Peacock, *Essential Grizzly*, xii.

14. For information about Pinchot and conservation, see Gifford Pinchot, "Ends and Means," 58–64; and Nash, *Rights of Nature*, 9. For conservation's relationship with Big Business, see Hays, "Conservation as Efficiency," 82–85.

15. The first anecdote is from Marsh, *Four Years in the Rockies*, 164–67. The tale about Lolo is from Space, *Lolo Trail*, 3. The last anecdote is from Moore, *Lochsa Story*, 266.

16. Moore, *Lochsa Story*, 266–69. For the Carlin Party, see Space, *Lolo Trail*, 48.

17. The majority of the paragraph is from Moore, *Lochsa Story*, 270–71. The numbers of sheep that grazed in the Bitterroots comes from Melquist, *Preliminary Survey*, HFP, ML. Information about salmon runs is from Stout, "Reintroduction of Grizzlies," 6. For information about the killing of the last bear in the Clearwater, see Space, *Clearwater Story*, 81.

18. Much of this paragraph, including Moore's quote, comes from Lapinski, *Grizzlies and Grizzled Old Men*, 85, 86. For information concerning his early childhood see Bud Moore's "Last of the Bitterroot Grizzly." For a brief biographical sketch of Moore, see Rob Chaney, "Forestry Pioneer, Conservation Icon Bud Moore Dies at Age 93," *Missoulian*, November 29, 2010.

19. Hays, *Beauty, Health, and Permanence*, 19, 35. For more detailed coverage of wildlife in the media, see Chris, *Watching Wildlife*.

20. The majority of the paragraph is from Hays, *Beauty, Health, and Permanence*, 22–24, 115–16.

21. Leopold, *Sand County Almanac*, 251, 262.

22. Leopold, *Sand County Almanac*, 138–39, 222.

23. For an overview of the switch from conservation to environmentalism, see Cawley, *Federal Land, Western Anger*, 16–21. For the ideological justifications for environmentalism, see Robert Marshall, "Wilderness," 123; and Wallace Stegner, "Meaning of Wilderness," 192–93. For the idea of predators as a symbol of wildness, see Mighetto, *Wild Animals*, 75. For Stewart Udall's thoughts see, Udall, "Prospects for the Land," 213.

24. Biel, *Do (Not) Feed the Bears*, 21–23.

25. The idea attributed to Mighetto is from Mighetto, *Wild Animals*, 91. The rest of the paragraph is from, Biel, *Do (Not) Feed the Bears*, 26, 41.

26. For visitation statistics, see "Yellowstone National Park Annual Park Visitation (All Years)," NPS STATS, accessed on June 6, 2013, https://irma.nps.gov/Stats/SSRSReports/Park%20specific %20reports/Annual%20park%20visitation%20%28all%20 years%29?Park=YELL.

27. Biel, *Do (Not) Feed the Bears*, 71. Also see U.S. Forest Service, "Smokey's Journey," accessed on June 6, 2013, http://www.smokey bear.com/vault.

28. Biel, *Do (Not) Feed the Bears*, 68–72. For information about the *True-Life Adventure* show, see Chris, *Watching Wildlife*, 30, 68.

29. Lapinski, *Grizzlies and Grizzled Old Men*, 109–20.

30. For information about the night of the grizzly, see Peacock and Peacock, *Essential Grizzly*, 120–21. For the drop in the bear population, see Lapinski, *Grizzlies and Grizzled Old Men*, 120.

31. S. Petersen, *Acting for Endangered Species*, 21–23.

32. Kohm, "The Act's History and Framework," 12–13, and S. Petersen, *Acting for Endangered Species*, 23–24.

33. Rohlf, *Endangered Species Act*, 19–20; and S. Petersen, *Acting for Endangered Species*, 25.

34. Nixon's statement is from, Richard M. Nixon and the Council on Environmental Quality, "Environmental Priorities for the 1970s," 250–51. The rise of the environmental movement, the significance of Earth Day in 1970, and the passage of the Marine Mammal Protection Act come from Rohlf, *Endangered Species Act*, 22–23. Facts

about the Marine Mammal Protect Act, CITES, and a call for stronger endangered species legislation is found in S. Petersen, *Acting for Endangered Species*, 27; and Nash, *The Rights of Nature*, 174–75.

35. Hays, *Beauty, Health, and Permanence*, 65–67.
36. The quotation from Nixon is from Richard Nixon, "51–Special Message to the Congress." Details concerning the passage of the act come from S. Petersen, *Acting for Endangered Species*, 27–30; and Kohm, "The Act's History and Framework," 15–16.
37. "To Pass S. 1983: A Bill Providing for the Conservation, Protection, and Propagation of Endangered Species," July 24, 1973, GovTrack, accessed on June 5, 2014, https://www.govtrack.us/congress /votes/93-1973/s313. "To Suspend the Rules and Pass H.R. 37, the Endangered and Threatened Species Act of 1973," September 18, 1973, GovTrack, accessed on June 5, 2014, https://www.govtrack.us /congress/votes/93-1973/h339.
38. Kohm, "The Act's History and Framework," 16–18.
39. Nash, *Rights of Nature*, 6–7.
40. Kellert, *Public Attitudes*, 21, 31, 101.
41. Kellert, *Public Attitudes*, 80, 133.
42. S. Petersen, *Acting for Endangered Species*, 30–35. Kohm, "The Act's History and Framework," 15.
43. Sutter, *Driven Wild*, 10. Rothman, *Greening of a Nation?*, 4.
44. The first statement is from Commoner, "Broader Context of the Environmental Movement," 244. The idea that the ESA was especially significant for predators is from Bixby, "Predator Conservation," 199.

2. Endangered Species, Environmental Politics

1. Egan, *Lasso the Wind*, 18–21.
2. Egan, *Lasso the Wind*, 23; and Rothman, *Greening of a Nation?*, 197–201.
3. Lapinski, *Grizzlies and Grizzled Old Men*, 121.
4. Kline, "Grizzly Bear Blues," 380–81. Also, see U.S. National Archives and Record Administration, "Amendment Listing the Grizzly Bear," 31734.
5. U.S. National Archives and Record Administration, "Amendment Listing the Grizzly Bear," 31734. For the Craighead's research see Lapinski, *Grizzlies and Grizzled Old Men*, 116. For the information

about grizzly bear reproduction see, Servheen, *Grizzly Bear 5-year Review*, 31–34.

6. Kline, "Grizzly Bear Blues," 383–84.

7. S. Petersen, *Acting for Endangered Species*, 43–44.

8. For more information, see Mann and Plummer, *Noah's Choice*, 164–69.The quotation from the Endangered Species Act comes from Endangered Species Act of 1973 (P.L. 93–205, 87 Stat. 884, Dec. 28, 1973; current version 16 U.S.C. 1531): 237. Kellert's quote is from Kellert, *Value of Life*, 164.

9. Mann and Plummer, *Noah's Choice*, 170–73.

10. Rothman, *Greening of a Nation?*, 4–5, 127–28.

11. For the quote, see Cawley, *Federal Land, Western Anger*, 69–70. Also, see Hays, *Beauty, Health, and Permanence*, 66–67.

12. The first part of the paragraph concerning the environmental movement is from Rothman, *Greening of a Nation?*, 5. Cawley, *Federal Land, Western Anger*, 71–72.

13. Cawley, *Federal Land, Western Anger*, 1–9.

14. Information about the New West/Old West dilemma is from Robb, Riebsame, and Gosnell, eds., *Atlas of the New West*, 95; and Cawley, *Federal Land, Western Anger*, 11.

15. Cawley, *Federal Land, Western Anger*, 2, 110–12.

16. Cawley, *Federal Land, Western Anger*, 162.

17. Rothman, *Greening of a Nation?*, 196–202. For more information, see Clearinghouse on Environmental Advocacy and Research, excerpts from *The Wise Use Movement*, box 9,HFP, ML.

18. S. Petersen, *Acting for Endangered Species*, 81–94. Also see, Yaffee, *Wisdom of the Spotted Owl*, 70, 73.

19. Yaffee, *Wisdom of the Spotted Owl*, xv, 114.

20. The local communities' responses in Oregon are found in Ted Gup, "Owl vs. Man," *Time*, June 25, 1990, 60. Also see Yaffee, *Wisdom of the Spotted Owl*, 127, 132.

21. Yaffee, *Wisdom of the Spotted Owl*, xviii–xix.

22. Constitution of the State of Montana, March 22, 1972, http://courts.mt.gov/content/library/docs/72constit.pdf.

23. For relationship between grizzlies and the spotted owl, see Bill Loftus, "GOP Committee Begins Fight Against Grizzly Area," *Lewiston Morning Tribune*, March 2, 1993; and Hank Fischer, "New Home for the Griz: Biologists Plan to Restore the Great Bear to the Lower 48's

Biggest Roadless Area," *Defenders Magazine* (Winter 1993–94), 19–20.

24. The first part of the paragraph, including the fact that many listed species were breeding in captivity, is from Bean, "Looking Back," 37–39. The middle of the paragraph, including ideas concerning megafauna, the act's ability to actually recover species, and the idea that it does not protect ecosystems is from Mann and Plummer, "Is Endangered Species Act in Danger?," 1257–58. The idea of decisions concerning the act being made on science is also from Bean, "Looking Back," 42. The final quotation is from Houck. "Reflections on the Endangered Species Act,"

25. O'Toole and Moskowitz, "Beyond the 100th Paradigm," 5.

26. Yaffee, *Wisdom of the Spotted Owl*, 73. Also see, Frank Clifford, "Environmental Movement Struggling as Clout Fades," *Los Angeles Times*, September 21, 1994.

27. For resentment of court decisions, see Wiebe, *Self-Rule*, 241. Also see, Frank Clifford, "Environmental Movement Struggling as Clout Fades," *Los Angeles Times*, September 21, 1994.

28. Rothman, *Greening of a Nation?*, 181–85.

29. For the drop in memberships, see Frank Clifford, "Environmental Movement Struggling as Clout Fades," *Los Angeles Times*, September 21, 1994. Also see O'Toole and Moskowitz, "Beyond the 100th Paradigm," 5.

30. For the immediate effects of the grizzlies' listing as threatened, see Peacock and Peacock, *Essential Grizzly*, 121. Also, see Kline, "Grizzly Bear Blues," 399–403. For information on the GBRP, see Servheen, *Grizzly Bear Recovery Plan*, ii.

31. Interagency Grizzly Bear Committee, *Looking Back*, 6, 12.

32. Knibb, *Grizzly Wars*, 28.

33. Melquist, *Preliminary Survey*, 6–9, 15–16, 31.

34. Bart B. Butterfield and John Almack, "Evaluation of Grizzly Bear Habitat in the Selway-Bitterroot Wilderness Area," University of Idaho, November 1985, ii–iii, 42, CAC, box 42, folder 10, BS.

35. Craig Groves, *A Compilation of Grizzly Bear Reports for Central and Northern Idaho*, Idaho Department of Fish and Game, February 1987.

36. Kunkel, Clark, and Servheen, *Remote Survey for Grizzly Bears*, 3.

37. For the first study, including types of grizzly food, see Dan Davis and Bart Butterfield, *The Bitterroot Grizzly Bear Evaluation Area: A Report to the Bitterroot Technical Review Team*, 1991, 32, 39, 40, HFP, ML. Christopher Servheen, Anthony Hamilton, Richard Knight, and Bruce McLellan, *Report of the Technical Review Team: Evaluation of the Bitterroot and North Cascades to Sustain Viable Grizzly Bear Populations*, Report to the Interagency Grizzly Bear Committee, 1991, 3, 6, HFP, ML. IGBC, "Statement of the IGBC on the North Cascades and Bitterroot," CAC, box 44, folder 23,BS.

3. Wolf Recovery Sets the Stage

1. For the capture and transportation of the wolves, see McNamee, *Return of the Wolf*, 62–77. For more about the wolves, see Fischer, *Wolf Wars*, 160.
2. For information about the scene at the arch, see McNamee, *Return of the Wolf*, 81–82; and Fischer, *Wolf Wars*, 161. The *High Country News* report can be found in Michael Milstein, "Happy Pack of Journalists Pursue Quarry," *High Country News*, February 6, 1995. For the lottery, see News Release, National Park Service, January 26, 1995, box 2, HFP, ML. For more about the holding pens, see U.S. Fish and Wildlife Service, "Draft Release #94-74," November, 1994, box 2, HFP, ML.
3. Bruce Babbitt's first quote is from Michael Milstein, "The Wolves Are Back, Big Time," *High Country News*, February 6, 1995. His second quote is from Fischer, *Wolf Wars*, 161.
4. McNamee, *Return of the Wolf*, 84–87.
5. Robinson, *Predatory Bureaucracy*, 14, 29; Fischer, *Wolf Wars*, 15.
6. Robinson, *Predatory Bureaucracy*, 30–31, 55, 77, 169; Fischer, *Wolf Wars*, 17–21.
7. Mighetto, *Wild Animals*, 79, 91; Fischer, *Wolf Wars*, 36–39.
8. U.S. Fish and Wildlife Service, *Final Environmental Impact Statement: Gray Wolves*, 1-4, 6–24. Fischer, *Wolf Wars*, 47–49.
9. Fischer, *Wolf Wars*, 49–52.
10. The majority of the paragraph is from Fischer, *Wolf Wars*, 52. For the last lines concerning Reagan, Watt, and the Sagebrush Rebellion, see Cawley, *Federal Land, Western Anger*, 12. (Larry Craig later gained national notoriety for a homosexual encounter in an air-

port bathroom in 2007 that prompted him not to seek reelection the following year.)

11. Endangered Species Act of 1973 (P.L. 93-205, 87 Stat. 884, Dec. 28, 1973; current version 16 U.S.C. 1531) Section 10(j).

12. "Northern Rocky Mountain Wolf Recovery Plan, 1987," in Schullery, *Yellowstone Wolf*, 240–46.

13. Fischer, *Wolf Wars*, 55–58, 66–67, 87. For the final arguments against wolf reintroduction, see Simpson, "Wyoming Wolves?," 8–11. Begley, "Return of the Native," 53. Simpson's second quote is from Alan Melnykovych, "On Wolves, Simpson's Just Full of Myth," *Casper Star Tribune*, June 23, 1986. The idea that wolf recovery was expensive is from Hank Fischer to Ralph Morgenweck, December 6, 1994, HFP, ML. For the idea of change, see Askins, "Releasing Wolves from Symbolism," 15.

14. Information on killings wolves in Yellowstone is from Cauble, "Return of the Native," 24–25. For the ecological information see, Matt Winters, "Dunkle Says NPS Wants Wolves to Undo Mistake," *Casper Star-Tribune*, November 8, 1987. Information on the relationship between wolves and bears is from Weaver, "Of Wolves and Bears," 27. For the quote, see Jean Heller, "Park Chief Wants Wolf Restored," box 17, HFP, ML.

15. Ted Kerasote, "Getting to Yes on the Wolf," *Sports Afield*, July 1989, 45–48. Leopold's quote is from Leopold, *Sand County Almanac*, 262.

16. Information about Baucus and Williams is from Fischer, *Wolf Wars*, 89–90. Information about Andrus is from Hays, *Beauty, Health, and Permanence*, 50; and Rocky Barker, "Cecil Andrus Knew How to Take a Stand," *High Country News*, February 20, 1995.

17. Sherry Devlin, "Wolf Champion Toasts Reintroduction," *Missoulian*, May 7, 1994. For information about Dunkle, see Andrew Melnykovych, "Wildlife Director Rolls Over for Sheep," *Casper Star-Tribune*, November 23, 1987.

18. Leopold's quote is from Leopold, *Sand County Almanac*, 262. For information about Mott, see Fischer, *Wolf Wars*, 88. For the survey, see McNaught, "Wolves in Yellowstone?," 519.

19. Fischer's quote is from Fischer, *Wolf Wars*, 66; also see Fischer, *Wolf Wars*, 88–91. For more about the campaign, see Defenders of Wildlife, "Wolf Education Project Wins National Award," News Release, May 20, 1986, box 2, HFP, ML. For information concerning

public support, see "Montanans Favor Reintroduction of Wolves, Poll Says," *Missoulian*, October 19, 1987; and Donna Gordon, "Survey Says Wyoming Residents Favor Returning Wolves to Park," *Casper Star-Times*, April 22, 1989. The idea of the media campaign as a turning point is from Sherry Devlin, "Wolf Champion Toasts Reintroduction," *Missoulian*, May 7, 1994.

20. For information about the Browning wolves, see Andrew Melnykovych, "Marauding Wolves May Be Killed in Montana," *Casper Star-Tribune*, August 19, 1987; and "Stockman Tells of Losses to Wolves," *Western Livestock Reporter*, December 17, 1987. For more on the compensation program, see Tom Blood, "Saving Wolves by Soothing Ranchers," *Wall Street Journal*, October 20, 1987. Because of the success of this program, Defenders of Wildlife decided to make this fund permanent, and for many years following wolf reintroduction, they continued to reimburse ranchers for lost livestock confirmed to have been killed by wolves.

21. For the numbers on the compensation program, see Defenders of Wildlife, "Payments to Ranchers from Defenders of Wildlife's Wolf Compensation Fund," Fact Sheet, box 2, HFP, ML. For Dunkle's resistance, see Bert Lindler, "Dunkle Not Ready to OK Wolves for Yellowstone," *Great Falls Tribune*, October 19, 1986; and Fischer, *Wolf Wars*, 96, 107. For more about national support, see U.S. Fish and Wildlife Service, *Final Environmental Impact Statement: Gray Wolves*, 6-32–6-33.

22. Orloff, *Environmental Impact Statement Process*, 6–7.

23. For Owens's bill, see JoAn Bjarko, "Utah Rep's Bill Would Force Wolf Decision in Park," *Bozeman Chronicle*, October 5, 1987. For information on the bills, see "Wolves and Congress: Proposed Legislation," in Schullery, *Yellowstone Wolf*, 248–53.

24. For information on McClure's bill, see "Wolves and Congress: Proposed Legislation," in Schullery, *Yellowstone Wolf*, 248–53. For more information on McClure, see Cauble, "Return of the Native," 25.

25. "Wolves and Congress: The Wolf Management Committee and the Funding of the EIS," in Schullery, *Yellowstone Wolf*, 254–59; and Fischer, *Wolf Wars*, 141.

26. Fischer, *Wolf Wars*, 144–48.

27. Fischer, *Wolf Wars*, 148–50, 155.

28. Fischer, *Wolf Wars*, 152–53, 157, and McNamee, *Return of the Wolf*, 81–87. If lawsuits had delayed reintroduction much longer, there is a good chance it would have never occurred. The conservative legislators who won seats during the Republican Party's sweep of the 1994 midterm elections took office just a few weeks later and would have likely blocked the reintroduction.

29. Askins's quote is from Begley, "Return of the Native," 53. Hank Fischer's quote is from Fischer, "Moving Past the Polarization," 1. Bruce Babbitt's quote is from Michael Milstein, "The Wolves Are Back, Big Time," *High Country News*, February 6, 1995. His second quote is from Fischer, *Wolf Wars*, 161. Also see Jones, "Way Out West," 33–34.

30. The killing of the wolf in Idaho is from "Wolves: Shoot First, Investigate Second," *Newsweek*, March 27, 1995, 60. For more about the investigation, see Dan Egan, "Autopsy Shows Wolf Didn't Kill Calf," *Idaho Post Register*, box 2, HFP, ML; and Dan Egan, "Feds Bare Own Claws Over Search," *Idaho Post Register*, box 2, HFP, ML.

4. The Advent of the ROOTS Coalition and the EIS

1. Dan Johnson to Interagency Grizzly Bear Committee, December 15, 1993, HFP, ML. For threats facing Idaho's timber industry, see James Peek, "Comments on Grizzly Bear Issues to the IGBC at Denver," December 15, 1993, HFP, ML. For Johnson's quote, see Werblow, "Timber & Wildlife Bear with Each Other," 20–21. For information about the IGBC meeting, see U.S. Fish and Wildlife Service, *Final Environmental Impact Statement: Grizzly Bear*, 5-3–5-4.

2. Interview with Hank Fischer, December 4, 2012, in possession of the author; Paul Larmer, "Soft-Path Approach to Saving Species," *High Country News*, May 15, 1995; and Hank Fischer, *Wolf Wars*, 183.

3. "Tom France," National Wildlife Federation, accessed on June 13, 2014, http://www.nwf.org/News-and-Magazines/Media-Center /Faces-of-nwf/Tom-France.aspx.

4. Tom Kenworthy, "Unlikely Alliance Finds Common Ground for Grizzlies," *Washington Post*, October 29, 1995.

5. Dant, "Making Wilderness Work," 239, 252, 260–61.

6. Dant, "Making Wilderness Work," 267–69.

7. U.S. Fish and Wildlife Service to John Mumma, June 1990, CAC, box 44, folder 23, BS. Also see multiple letters to Cecil Andrus housed in CAC, box 44, folder 23, BS.

8. For the quote, see Julie Titone, "Grizzly's Status Threatened by Modest Recovery," *Spokane Spokesman-Review*, June 7, 1993. For the potential lawsuit, see Sierra Club Legal Defense Fund, "Environmentalists Serve Notice of Lawsuit to Protect Critically Endangered Grizzly Bear Population," News Release, March 30, 1993, Larry LaRocco Papers, box 29, folder 3, BS.

9. For the initial timeline, see U.S. Fish and Wildlife Service, *Final Environmental Impact Statement: Grizzly Bear*, 5-3-5-4. For the first information concerning Wakkinen, see Hank Fischer, "Idaho Could Be Sending Wrong Message on Grizzlies," *Post Register*, February 28, 1993; and U.S. Fish and Wildlife Service, *Final Environmental Impact Statement: Grizzly Bear*, 5-3-5-4.

10. For information about the CIG, see "Mission Statement: Bitterroot Grizzly Bear Citizen's Involvement Group," and Bitterroot Grizzly Bear Recovery Citizen's Involvement Group, Meeting Minutes, August 1, 1992, HFP, ML. Also see Wayne Wakkinen to Hank Fischer, August 27, 1992, and Wayne Wakkinen [to Hank Fischer?], August 20, 1992, HFP, ML. For information about the cancelled meeting, see Bill Loftus, "Planned Grizzly Restoration Meeting Put in Hibernation," *Lewiston Morning Tribune*.

11. For the first information in the paragraph, see U.S. Fish and Wildlife Service, *Final Environmental Impact Statement: Grizzly Bear*, 5-3-5-4. For Servheen's quote and information about the meetings, see Bill Loftus, "Recovery Plan Is Due Out Next Spring," *Lewiston Tribune*, September 25, 1992. Also see Northwest Ecosystem Subcommittee Meeting, "Bitterroot Ecosystem Summary," November 17, 1992, HFP, ML. For Britt's quote, see Bill Loftus, "Biologists, Timber Workers Debate Grizzly Reintroduction," *Lewiston Morning Tribune*, September 28, 1993.

12. For information on the bill and Conley's statement, see, Rocky Barker, "Bill Would Restrict F&G Control over Bear Management," *Post Register*, February 23, 1993. For Fischer's statements, see Hank Fischer, "Idaho Could Be Sending Wrong Message on Grizzlies," *Post Register*, February 28, 1993. Also see Stephen Stuebner, "Stream Flow Bill Passed by House," *Post Register*, March 3, 1993. The last quote is from "Idaho Doesn't Need Grizzly Committee," *Idaho Statesman*, March 4, 1993.

13. Larry Craig's quote is from Larry Craig to Chris Servheen, October 20, 1993, HFP, ML. Andrus's quote is from Bill Loftus, "GOP Com-

mittee Begins Fight against Grizzly Area," *Lewiston Morning Tribune*, March 2, 1993. For the idea that grizzlies are an excuse to create wilderness, see Cloan W. McNall, "Grizzly's Re-introduction Threatens Resource Uses," *Clearwater Tribune*, March 18, 1993. The quote about fear is from Greg Billups, "Fear in the Forest," *Lewiston Morning Tribune*, March 14, 1993. Andrus's quote is from "Gov. Andrus Cool to Grizzly Bear Recovery Plans," *Lewiston Morning Tribune*, May 16, 1993. The final quote and sentence is from Bill Loftus, "Bears Find Few Friends at Orofino," *Lewiston Tribune*, August 24, 1993.

14. Backcountry Horsemen of Idaho, "Bitterroot Grizzly Bear Recovery Position Paper"; Clearwater County Board of Commissioners, "Grizzly Bear Proposal"; and Dan Johnson to Interagency Grizzly Bear Committee, December 15, 1993, CAC, box 42, folder 10, BS.

15. For information about grizzly habitat, see Servheen, *Grizzly Bear Recovery Plan*, 3, 7. For the first objection see, John E. Burns to Chris Servheen, June 11, 1993, HFP, ML. For the second objection see, Stephen K. Kelly to Chris Servheen, June 14, 1993, HFP, ML. For the final objection, see Orville L. Daniels to Chris Servheen, June 11, 1993, HFP, ML. Bitterroot Citizen's Involvement Group, Meeting Minutes, June 19, 1993, HFP, ML.

16. Servheen, *Grizzly Bear Recovery Plan*, 11–12, 23–24.

17. For the Bitterroot Subcommittee's decision to let the EIS handle the controversial issues, see Adam Ruben, Memo: Selway-Bitterroot Ecosystem Subcommittee Meeting, November 16, 1993, and Herb Pollard, Minutes: Bitterroot Ecosystem Subcommittee Meeting, November 16, 1993, HFP, ML. For Ruben's objections, see Adam Ruben to Bitterroot Griz Activists, December 8, 1993, and Adam Ruben to Herb Pollard, December 8, 1993, HFP, ML.

18. Christopher Servheen, "Grizzly Bear Recovery Plan Supplement: Bitterroot Ecosystem Recovery Plan Chapter," September 11, 1996, HFP, ML. For information about the Cascades, see Knibb, *Grizzly Wars*, 128.

19. For Americans' history with pragmatic politics, see Kloppenberg, "An Old Name for Some New Ways of Thinking?," 100. Hibbard and Madsen, "Environmental Resistance," 706–8. Also see Conley and Moote, "Evaluating Collaborative Natural Resource

Management," 372; and Kagan, "Political and Legal Obstacles to Collaborative Ecosystem Planning," 873. For the influence of the Sagebrush Rebellion, see Cawley, *Federal Land, Western Anger*, 153–54.

20. Hibbard and Madsen, "Environmental Resistance," 711.

21. Cestero, *Beyond the Hundredth Meeting*, 21–25, 30–31, 49.

22. For funding requests, see Hank Fischer and Tom France to Neal Sigmon, September 28, 1993, and Hank Fischer and Tom France to Debbie Weatherly, September 28, 1993, HFP, ML. Also see Tom France and Michael Roy to Chris Servheen, October 18, 1993, HFP, ML. For information about grizzly bear breeding and effects of human-caused mortality see, Servheen, *Grizzly Bear Recovery Plan*, 4–7, 28–29.

23. For Defenders of Wildlife's grassroots campaign, see Lisa Lombardi to Hank Fischer, September 21, 1994, and Hank Fischer, Memorandum: Bitterroot Grizzly Bear Restoration, April 26, 1994, HFP, ML. Fischer's quote is from Dan Whipple, "Groups Cooperate on Idaho Grizzly Plan," *Star-Tribune*, January 18, 1996.

24. Dan Johnson to Interagency Grizzly Bear Committee, December 15, 1993, HFP, ML.

25. See James Peek, "Comments on Grizzly Bear Issues to the Interagency Grizzly Bear Committee at Denver," December 15, 1993, HFP, ML; "Orofino Hearing Comments," Grizzly Bear Management Oversight Committee, August 23, 1993; "Idaho Lawmaker Backs Bitterroot Grizzly Bear Plan," *Missoulian*, January 9, 1994; and Idaho Grizzly Bear Oversight Committee, Position Paper, February 4, 1994, HFP, ML.

26. For ROOTS commitment to grizzly recovery see Dan Johnson, Gerry Snyder, Greg Gollberg, James Peek, and Bill Wall to Chris Servheen, May 13, 1994, HFP, ML. For information about their second meeting, see Dan Johnson to Hank Fischer, April 6, 1994, HFP, ML.

27. For information about the NWF's brochure, see Michael Roy to Dan Johnson, May 16, 1994, HFP, ML. For the second meeting's agenda, see Hank Fischer to Dan Johnson, June 6, 1994, HFP, ML. For more about the meeting, see the interview with Hank Fischer, March 2, 2012, in possession of the author.

28. Information about attendance to the meeting and invitations to the group is from Hank Fischer to Mike Roy, Lisa Lombardi, John McCarthy, Louisa Willcox, and Dennis Baird, June 13, 1994, HFP, ML.

29. Unfortunately, Seth Diamond died in a plane crash in the summer of 1996. His tragic death dealt a severe blow to the coalition's momentum. The IFIA's initial resistance is from IFIA, "Hot Sheet: Grizzly Bear Recovery Plan Needs Comments," CAC, box 44, folder 23, BS. IFIA's involvement is from Chuck Lobdell to Hank Fischer, Seth Diamond, Tom France, and Bill Mulligan, November 29, 1994; Seth Diamond to Bitterroot Operators, September 28, 1994; and Seth Diamond to Idaho operators, November 12, 1994, all from the Hank Fischer Papers, ML. Information about Seth Diamond and Diamond's death is from Ed Marston, "Six Views of the Grizzly Bear Controversy," *High Country News*, May 13, 1996; and Jim Riley, "Testimony for Oversight Hearing on Reintroduction of Grizzly Bear in the Public Domain National Forests," June 12, 1997, 146, HFP, ML. Sherry Jones, "Montana Has Truly Lost a Friend," *Missoulian*, July 28, 1996.

30. For ROOTS's initial position, see Dan Johnson to IGBC, December 15, 1993, HFP, ML. For their revised position, see ROOTS, "Draft Revised Grizzly Bear Position Paper," September 20, 1994, HFP, ML.

31. ROOTS, "Draft Revised Grizzly Bear Position Paper," September 20, 1994, HFP, ML.

32. For initial lobbying for EIS funding, see Rodger Schlickeisen to Mollie Beattie, May 10, 1994, HFP, ML. For information about support from LaRocco and Williams, see Larry LaRocco and Pat Williams to Sidney Yates, June 1, 1994, HFP, ML. Williams's position on Cabinet grizzlies is from a Pat Williams letter, May 4, 1988, box 11, folder 3, Pat Williams Papers, ML. For support for the EIS from other organizations, see James M. Peek to Larry LaRocco, May 17, 1994, HFP, ML; C. Richard Clough to Michael Roy, May 18, 1994, HFP, ML; and Jerry Conley, "Idaho Fish & Game Position Paper, Grizzly Bear Recovery," HFP, ML. Mulligan's quote is from Hank Fischer, "Bears and the Bitterroot," 18.

33. For Servheen's estimates, see Christopher Servheen, "Memorandum: Draft Budget for Bitterroot Ecosystem EIS/NEPA Process," February 17, 1994, HFP, ML. Information about the Federal Register

is from U.S. National Archives and Record Administration, "Reintroduction of Grizzly Bears," 2399–400. Credit for the grizzly EIS being funded being given to the ROOTS groups is from Fischer, "Moving Past the Polarization," 6.

34. The statistics about the cost are from Fischer, "Moving Past Polarization," 10. Quote is from Rodger Schlickeisen to Mollie Beattie, May 10, 1994, HFP, ML.

5. Environmental Resistance

1. For information about Bader and the AWR, see the interview with Mike Bader, December 5, 2012, in possession of the author. For information about NREPA, see Alliance for the Wild Rockies, "NREPA," accessed on July 19, 2013, http://www.wildrockiesalliance .org/issues/nrepa.

2. For information about the AWR taking the lead on the issue, see the interview with Mike Bader, December 5, 2012, in possession of the author.

3. Interview with Mike Bader, December 5, 2012, in possession of the author.

4. Interview with Mike Bader, December 5, 2012.

5. For the appointment of John Weaver, see Chris Servheen to Hank Fischer, January 30, 1995, HFP, ML. For information about involvement in the recovery team and the February meeting see, Herb Pollard, Minutes for Bitterroot Ecosystem Management Committee Meeting, February 2, 1995, HFP, ML. For more about the ROOTS coalition pushing the faster timeline, see Michael Roy, Dan Johnson, Hank Fischer, and Seth Diamond to Herb Pollard, January 31, 1995, HFP, ML.

6. Knibb, *Grizzly Wars*, 129–31.

7. U.S. Fish and Wildlife Service, *Final Environmental Impact Statement: Grizzly Bear*, 5-4-5-5; and Pollard, Minutes for Bitterroot Ecosystem Management Committee Meeting, February 2, 1995, HFP, ML. Also see Hank Fischer, Memorandum: Meeting to Discuss Bitterroot Grizzly Restoration, October 17, 1994, HFP, ML.

8. National Wildlife Federation, Resource Organization on Timber Supply, Defenders of Wildlife, Intermountain Forest Industry Association, "Working Draft: Bitterroot Grizzly Recovery Principles," January 24, 1995, HFP, ML.

9. The quote is from Hank Fischer, "Bears and the Bitterroot," 25.

10. For Fischer and France's request for support, see letter by Hank Fischer and Tom France, March 2, 1995; and Hank Fischer and Tom France to Louisa Willcox, March 7, 1995, HFP, ML.

11. For Baird's comment, see Dennis Baird to Tom France, July 14, 1995, HFP, ML. For other reactions, see John McCarthy to Tom France and Hank Fischer, March 14, 1995; and Dan Pletscher to Hank Fischer, March 13, 1995, HFP, ML.

12. For the response, see Brian Peck, Louisa Willcox, John McCarthy, Michael Scott, and Larry Mehlhaff to Tom France and Hank Fischer, March 2, 1995, HFP, ML. Reid's quote is from Ken Miller, "Environmentalists Split over Terms for Grizzly Return," *Idaho Statesman*, July 21, 1995.

13. For these charges against collaboration see Petersen et al., "Conservation and the Myth of Consensus," 762–64; Conley and Moote, "Evaluating Collaborative Natural Resource Management," 373; Hibbard and Madsen, "Environmental Resistance," 708–11; and Moote and McClaren, "Viewpoint," 479–80.

14. The quotations are from, "An Educational Alert from the Ecology Center and Alliance for the Wild Rockies," *North Rockies Network News*, June 27, 1995. For the timeline of the CBA, see the interview with Mike Bader, December 5, 2012, in possession of the author.

15. Bader, Garrity, and Bechtold, *Conservation Biology Alternative for Grizzly Bear Population Restoration*, 4, 6, 12, 19.

16. Michael Roy to Hank Fischer, December 18, 1995, HFP, ML.

17. Michael Roy to Hank Fischer, December 18, 1995, HFP, ML. For road densities, see Hank Fischer and Tom France to Mike Phillips, May 1, 1998, HFP, ML; and Servheen, *Grizzly Bear Recovery Plan*, 147. For acres logged, see Steve Thompson, "Impacts of Current and Projected National Forest Management on Grizzly Bear Recovery in the Bitterroot Ecosystem," National Wildlife Federation and Defenders of Wildlife, March 2000, HFP, ML. For France's estimation, see Tom France to Don Judge, June 21, 1996, Minette Johnson Papers, Defenders of Wildlife, Missoula MT. For road building, see Sherry Devlin, "Report Says Bitterroots Ideal for Griz," *Missoulian*, April 6, 2000.

18. The citation about linkage corridors is from Hank Fischer and Mike Roy to Dominick DellaSala and Paul Paquet, April 23, 1996; and Rodger Schlickeisen and Mark Van Putten to Bill Meadows, March

19, 1997, HFP, ML. For Bader's quote, see the interview with Mike Bader, December 5, 2012, in possession of the author.

19. Nadeau's quote is from Dan Gallagher, "Education Now Key to Grizzly Plan," *Clearwater Register*, January 16, 1996. For more on Roy's opinion on the CBA, see Mike Roy to Reed Noss, HFP, ML. For France's quote, see Tom France to Don Judge, June 21, 1996, Minette Johnson Papers, Defenders of Wildlife, Missoula MT.

20. Solomon is not using "conservationist" as I defined the term in the introduction. His use of the word reflects the reality that the term "environmentalist" had become synonymous with radicalism and was thus politically indefensible.

21. The idea that the CBA was meant as merely a foil is from the interview with Mike Bader, December 5, 2012, in possession of the author. Ideas about differences in breeding between wolves and bears is from George Kenneth to Mike Roy, January 30, 1996, HFP, ML. Olsen's quote is from Justin Smith, "Groups Want Bears Protected," *Missoulian*, July 16, 1995. Success of experimental populations is from Environmental Defense Fund to Chris Servheen, November 19, 1997, HFP, ML.

22. Black Footed Ferret Recovery Implementation Team, "Black Footed Ferret Recovery Program," accessed on November 17, 2013, http://www.blackfootedferret.org/. Defenders of Wildlife, "Fact Sheet: California Condor," accessed on November 17, 2013, http://www.defenders.org/california-condor/basic-facts.

23. U.S. Fish and Wildlife Service, "Species Profile: Whooping Crane," accessed on Nov. 17, 2013, http://ecos.fws.gov/speciesProfile/profile/speciesProfile.action?spcode=b003; and International Crane Foundation, "Historic Whooping Crane Numbers," accessed on Nov. 17, 2013, https://www.savingcranes.org/images/stories/site_images/conservation/whooping_crane/pdfs/historic_wc_numbers.pdf.

24. Not wanting Idaho and Montana to have control over management is from Andy Stahl to Hank Fischer and Tom France, January 29, 1996, HFP, ML. The belief that the citizen management committee will be partisan is from Marty Berghoffen to Defenders of Wildlife and National Wildlife Federation, HFP, ML. Brian Peck's opposition is from Brian Peck to Hank Fischer and Mike Roy, March 7, 1996, HFP, ML. The idea that grizzlies were still in the Bitterroots is from "Grizzly Bear Population Recovery," HFP, ML.

25. "Endangered Species Act, Rule 10(j) Reintroduction of Grizzly Bears into the Bitterroot Grizzly Bear Recovery Zone, Revised Draft," November 28, 1995. Document can be accessed at www.fws .gov/endangered/laws-policies/section-10.html.

26. Fischer's quote is from Hank Fischer, Tom France, and Mike Roy, Memorandum: Update on Bitterroot Grizzly Restoration, January 4, 1996, HFP, ML. Diamond's observation is from Seth Diamond to Bob Ferris, March 16, 1996, HFP, ML. Roy's comment is from Mike Roy to Reed Noss, HFP, ML.

27. Most of the paragraph comes from an interview with Minette Johnson, September 23, 2013, in possession of the author. Also see Hank Fischer, Memo, April 1996, HFP, ML.

28. The importance of the Clinton administration's support is from Hank Fischer, Memorandum: Appropriations for Bitterroot Grizzly EIS, May 13, 1994, HFP, ML. For Clinton and the environmental movement, see Patterson, *Restless Giant*, 349, 379. The effect of the reduced costs can be found in ROOTS to George Frampton, November 12, 1996, HFP, ML. The fact that the USFWS considered not releasing the EIS is from Rodger Schlickeisen and Mark Van Putten to Bruce Babbitt, January 8, 1997, HFP, ML. Babbitt's support is from Rodger Schlickeisen and Mark Van Putten to Bruce Babbitt, March 13, 1997, HFP, ML.

29. Interviews by the author with Mike Bader, December 5, 2012; and with Hank Fischer, December 4, 2012, both in possession of the author.

6. Ethical Controversies and the Draft EIS

1. The first two quotations from Dennis Palmer and Robert Norton are taken from Ginny Merriam, "No Griz, No Way: Bitterrooters Don't Want Bears in Their Backyard," *Missoulian*, July 7, 1995. This meeting was held two years prior to the other meetings, before the U.S. Fish and Wildlife Service released the draft environmental impact statement. The anecdote about the women reading the pathology report is from Jane Rider, "Grizzly Options: Bitterrooters Split 50–50 on Bear Reintroduction," *Missoulian*, October 2, 1997. The comment about rattlesnakes and water moccasins comes from Sherry Devlin, "Most Want More Area for Bears," *Missoulian*, October 3, 1997. The incident of the man holding up his daughter can be found in "Bear-

ing With It: Bitterrooters Grapple with the Government's Decision to Bring Grizzly Bears Back to Western Montana," *Missoula Independent*, November 23, 2000. Finally, the quote from *High Country News* is from Louise Wagenknecht, "Grizzlies and the Male Animal," *High Country News*, October 27, 1997.

2. The first quotations from the Boise meeting are from "Grizzly Reintroduction Opponents, Proponents Square Off at Hearing," *Missoulian*, October 6, 1997. The account of the Missoula meeting is from Devlin, "Most Want More Area for Bears," *Missoulian*, October 3, 1997; and the account from Hamilton is from Rider, "Grizzly Options: Bitterrooters Split 50–50 on Bear Reintroduction," *Missoulian*, October 2, 1997. Christopher Servheen's quotation came from Sherry Devlin, "Grizzlies OK'd for Bitterroot," *Missoulian*, November 17, 2000.

3. Support from LaRocco and Williams can be found in Larry LaRocco and Pat Williams to Sidney Yates, June 1, 1994, HFP, ML. ROOTS's strategy is described in Hank Fischer to Herb Pollard, January 24, 1995, HFP, ML.

4. Rothman, *Greening of a Nation?*, 203. Patterson, *Restless Giant*, 324–25, 344–45. Information about Gingrich and the Republican agenda is from Barnes, "Revenge of the Squares," 13; and Gloria Borger, "Welcome to Gingrich Nation," *U.S. News & World Report*, November 21, 1994, 44–46.

5. Michael Janofsky, "Counting the Vote: The Montana Governor," *New York Times*, November 21, 2000; and Richard L. Berke, "Man in the News; Party Chairman, Presidential Friend—Marc Racicot," *New York Times*, December 6, 2001. For information about Racicot's hunters' citizen advisory council, see Guynn and Landry, "Case Study of Citizen Participation," 392–98.

6. For Racicot's testimony, see Marc Racicot, S. 1180 The Endangered Species Recovery Act of 1997–Oral Testimony, box 15, HFP, ML. For Racicot's support of ROOTS, see Bitterroot Ecosystem Subcommittee, Meeting Minutes, May 11, 1995, HFP, ML. For Racicot's quote, see Justin Smith, "Racicot Makes the Rounds," *Missoulian*, September 14, 1995.

7. For Batt's initial support, see Bitterroot Ecosystem Subcommittee, Meeting Minutes, May 11, 1995, HFP, ML. For his quotes and change of heart, see "Grizzly Expense: Idaho Fears Bears' bite on Economy," *Missoulian*, October 8, 1995.

8. For Chenoweth's initial interest, see Bitterroot Ecosystem Subcommittee, Meeting Minutes, May 11, 1995, HFP, ML. For her later change of heart, see Helen Chenoweth, "Don't Bring Back the Bears," *Washington Times*, December 12, 1995. For reactions to Chenoweth's position, see Dan Popkey, "Which Chicago Bears?" *Idaho Statesman*, December 19, 1995, and Allen Walker letter to the editor in the *Clearwater Progress*, October 26, 1995, HFP, ML. Also, see Christopher Georges, "House GOP Freshmen Are Facing Bumpy Roads in Election Drive as Counterrevolution Spreads," *Wall Street Journal*, July 24, 1996, and O'Connell, "Wise Use Groups Lobby Against Species Act,", 15.

9. Kirk Johnson, "Bush's Interior Nominee: Comfort in Consensus," *New York Times*, April 8, 2006. David Stout, "Idaho Governor Chosen to Be Interior Secretary," *New York Times*, March 16, 2006.

10. For the meeting with Kempthorne, see Minutes: Bitterroot Ecosystem Subcommittee Meeting, May 11, 1995, HFP, ML. For Kempthorne's bills see O'Connell, "Wise Use Groups Lobby Against Species Act," 15; and Robert Pear, "Source of State Power Is Pulled from Ashes," *New York Times*, April 16, 1995. Also see "Senators Kempthorne, Chafee Poised to Introduce 'Species Extinction Bill,'" *Endangered Species Northwest* (Northwest Ecosystem Alliance), issue 20, February 4, 1997, box 15, HFP, ML.

11. Michael Janofsky, "Counting the Vote: The Montana Governor," *New York Times*, November 21, 2000; and Richard L. Berke, "Man in the News; Party Chairman, Presidential Friend—Marc Racicot," *New York Times*, December 6, 2001.

12. Interview with Minette Johnson Glaser, September 23, 2013, in possession of the author.

13. For a list of the meetings held, see U.S. Fish and Wildlife Service, *Final Environmental Impact Statement: Grizzly Bear*, 5–5. For the *Missoulian* editorial, see "A Better Road to Recovery," *Missoulian*, July 5, 1995. For Diamond's quote, see Sherry Devlin, "Return of the Grizzly," *Missoulian*, July 4, 1995. The last three quotes are from Jonathan Brinckman, "Disparate Sides Join to Suggest Bear Plan," *Idaho Statesman*, July 21, 1995; "Refreshingly Open Minds," *Bozeman Daily Chronicle*, July 25, 1995; and J. F., "Grizzlies Get What Owls, Salmon Did Not: Consensus," *Lewiston Morning Tribune*, August 23, 1995. The last idea is from "On Resource Issues, Local Control Makes Sense," *Bozeman Daily Chronicle*, August 21 1997.

14. The stock growers' announcement is from the *Stockgrower Newsletter*, Montana Stockgrowers Association, June 23, 1995. The information from the Challis meeting is from Minette Johnson to Hank Fischer and Tom France, October 9, 1997, HFP, ML; and interview with Minette Johnson Glaser, in possession of the author.

15. The quotations and information from the meeting in Hamilton is from Merriam, "No Griz, No Way: Bitterrooters Don't Want Bears in Their Backyard," *Missoulian*, July 7, 1995.

16. Interview with Claire Kelly, October 23, 2013, in possession of the author.

17. The reference to people who did not believe they were extremists is from George Bursell, "Naturally, WE'RE wrong . . . ," *Missoulian*, July 16, 1995; and Dwight Finney, "Don't Dismiss Public Opinion," *Missoulian*, July 16, 1995. For the Forest Service employee's objections, see Ed Bloedel, "To the USFWS about Bears," *Ravalli Republic*, July 25, 1995. For fear of restrictions, see Pam Bradshaw, "Horsemen Oppose Grizzly's Return," *Ravalli Republic*, July 27, 1995. The Bitterrooter who feared for the safety of the bears came from Doris Milner, "Letters: Don't Experiment with Grizzlies," *Missoulian*, October 3, 1997.

18. For the initial plan, see Don Schwennesen, "Cabinets May Lose Grizzlies without Transplant," *Missoulian*, January 29, 1988. For Vincent's quote, see "New Cabinet Grizzly Plan Pleases Coalition," *Missoulian*, July 7, 1988. For the information about Lincoln County's opposition, see Shaun Tatarka, "Bears Have Affected Lincoln County," *Ravalli Republic*, July 28, 1995.

19. For information about the survey, see Sherry Devlin, "Grizzlies Belong, Residents Tell Survey," *Missoulian*, August 27, 1995. For Racicot's support and quote, see Marc Racicot to Steven Benedict, August 1, 1995, HFP, ML; and Marc Racicot, "Remarks to Montana Wood Products Association, Polson MT," August 24, 1995, HFP, ML. For Baucus' support, see Max Baucus to Bruce Babbitt, September 14, 1995, HFP, ML.

20. Conrad Burns's disapproval is described in Conrad Burns to Bruce Babbitt, October 23, 1995, HFP, ML. For the revised plan, see "Endangered Species Act, Rule 10(j) Reintroduction of Grizzly Bears into the Bitterroot Grizzly Bear Recovery Zone, Revised Draft," November 28, 1995. For responses to Marc Racicot's comments, see Wayne to Marc

Racicot, February 15, 1996, HFP, ML. For ROOTS's acceptance of the
Oversight Committee's suggestions, see letter of Mike Roy, Seth Diamond, Hank Fischer, and Tom France to Idaho Grizzly Oversight Committee, December 19, 1995, HFP, ML.

21. For the support of the Clearwater Resource Council, see Ron Hoff to
John Weaver, August 17, 1995, HFP, ML. For support from the IOGA,
see Grant Simonds to Idaho Grizzly Bear Management Oversight
Committee, December 19, 1995, HFP, ML. For information about
the support of the Oversight Committee, see James Peek to Seth
Diamond, Hank Fischer, Tom France, Dan Johnson, and Bill Mulligan, September 20, 1995, HFP, ML. For Racicot's support, see Hank
Fischer, Tom France, and Mike Roy, Memorandum: Update on Bitterroot Grizzly Restoration, January 4, 1996, HFP, ML. For Moore's
support, see Bud Moore to Doris Milner, August 5, 1997, box 52,
folder 8, Doris and Kelsey C. Milner Papers, ML.

22. Information about Racicot's meeting in Hamilton and the petition
is from Justin Smith, "Racicot Makes the Rounds," *Missoulian*, September 14, 1995. For information about the CAG group, see the interview with Claire Kelly, October 23, 2013, in possession of author.
For the poll, see Carlotta Grandstaff, "Ban the Bears," *Stevensville
Star*, January 21, 1996. IDFG's position is from Idaho Department
of Fish and Game, "Position Statement on Reintroduction of Grizzly Bears," November 14, 1995, HFP, ML. For Idaho's resolution, see
Idaho State Legislature, "House Joint Memorial no. 6," February 13,
1996, http://www.state.id.us:80/oasis/hjm006.htm., HFP, ML.

23. Most of the paragraph is from Hank Fischer to Tom France, Jim
Riley, Bill Mulligan, and Dave Halley, September 16, 1996, HFP, ML.
For more information on Chenoweth and Batt, see Phillip E. Batt to
Helen Chenoweth," January 25, 1996; Hank Fischer and Mike Roy
to Dominick DellaSala and Paul Paquet, April 23, 1996, HFP, ML. For
more on Kempthorne, see Hank Fischer and Tom France to Seth
Diamond, Dan Johnson, and Bill Mulligan, April 30, 1996, HFP, ML;
and Ed Marston, "Bringing Back Grizzlies Splits Environmentalists,"
High Country News, May 13, 1996.

24. Hank Fischer to Tom France, Jim Riley, Bill Mulligan, and Dave
Halley, September 16, 1996, HFP, ML.

25. The effect of the reduced costs is from ROOTS to George Frampton,
November 12, 1996, HFP, ML. The fact that the USFWS considered

not releasing the EIS can be found in Rodger Schlickeisen and Mark Van Putten to Bruce Babbitt, January 8, 1997, HFP, ML.

26. Nadeau's statement is from Gallagher, "Education Now Key to Grizzly Plan," *Clearwater Register*, January 16, 1996. Information about the sanitation survey is from Mike Roy to Jim Riley, August 26, 1996, and Nancy Heideman to Hank Fischer, August 22, 1996, HFP, ML. For the education plan, see "Endangered Species Act, Rule 10(j) Reintroduction of Grizzly Bears into the Bitter Grizzly Bear Recovery Area," Revised Draft, May 20, 1996.

27. Babbitt's support is from Rodger Schlickeisen and Mark Van Putten to Bruce Babbitt, March 13, 1997, HFP, ML. For Idaho's resolution, see Idaho State Legislature, "House Joint Memorial no. 2," First Regular Session, 1997, HFP, ML. Idaho Fish and Game's vote is from John Rosapepe and Steve Stuebner, "Idaho Says No Grizzlies," *High Country News*, February 17, 1997. For the spring meetings, see Steve Mealy, "Testimony for Oversight Hearing on Reintroduction of Grizzly Bear in the Public Domain National Forests," June 12, 1997, HFP, ML. The survey is from Idaho Consulting International, "Idaho Voter Opinion Poll," February 6, 1997, HFP, ML. For information about Kempthorne and Crapo, see Dirk Kempthorne, Michael Crapo, Helen Chenoweth, and Larry Craig to Bruce Babbitt, May 15, 1997, HFP, ML. For information about Burns, see Conrad Burns, "Burns Opposes Grizzly Reintroductions," News Release, June 27, 1997, HFP, ML.

28. David Scott, "Idaho Governor Chosen to Be Interior Secretary," *New York Times*, March 16, 2000; Kirk Johnson, "Bush's Interior Nominee: Comfort in Consensus," *New York Times*, April 8, 2006; and Michael Janofsky, "Idaho Governor Selected to Lead Interior Dept.," *New York Times*, March 17, 2006.

29. For the winter 1996 survey, see "Public Knowledge, Opinion, and Attitudes toward Endangered Species Reintroduction," 11. For the spring survey, see "Public Opinions and Attitudes toward Reintroducing Grizzly Bears to the Selway-Bitterroot Wilderness Area of Idaho and Montana," 7, 13. For articles from out of state, see "Bear Essentials: Blurring Battle Lines," *Arizona Republic*, May 25, 1997; Ken Olsen, "Survey: Bring grizzly bears back to Idaho wilderness," *Anchorage Daily News*, June 22, 1997, "Poll shows support for grizzly reintroduction," *Roseburg OR News-Review*, June 22, 1997; Jim Rob-

bins, "Plan to Repopulate Grizzlies Gains Support," *New York Times*, April 27, 1997; and Tom Kenworthy, "Politics Imperils Uncommon Alliance's Plan to Find Grizzlies a Home," *Washington Post*, October 12, 1997.

30. Reid's quote is from Ken Miller, "Environmentalists Split over Terms for Grizzly Return," Gannett News Service, July 21, 1995. Craighead's quote is from John J. Craighead, Jay Sumner, and John Mitchell, *Population Recovery in the Bitterroot/Selway Wilderness*, August 18, 1995, HFP, ML. The final quote is from "Grizzly Bear Population Recovery," HFP, ML. The survey data is from Mark D. Duda and Kira C. Young, "The Public and Grizzly Bear Reintroduction in the Bitterroot Mountains of Central Idaho," July 1995, Responsive Management Report, Harrisonburg VA, HFP, ML. The final quote is from "Grizzly Bear Population Recovery."

31. Carlson's quote is from Rita Carlson, "Testimony for Oversight Hearing on Reintroduction of Grizzly Bear in the Public Domain National Forests," June 12, 1997, 125, commdocs.house.gov/commit tee/resources/hii44273.000/hii44273_0f.htm. Most quotes come from Velado, "Grizzly Bear Reintroduction to the Bitterroot Eco-system," 69, 60, 63. The quote calling it a waste of money is from Mary Hopkin, "Grizzly Bear Committee Supporters Outline Efforts," Superior Clipping Service, Glendive MT, May 7, 1996. The survey data is from Mark D. Duda and Kira C. Young, "The Public and Grizzly Bear Reintroduction in the Bitterroot Mountains of Central Idaho," July 1995, Responsive Management Report, Harrison-burg VA, HFP, ML. The last quote is from Michael Koeppen to Hank Fischer, May 11, 1994, HFP, ML.

32. Interview with Claire Kelly, October 23, 2013, in possession of the author.

33. U.S. Fish and Wildlife Service, *Grizzly Bear Recovery*, 7.

34. U.S. Fish and Wildlife Service, *Grizzly Bear Recovery*, 7, 12–13, 18–19.

35. Information about the meetings is from U.S. Fish and Wildlife Service, *Final Environmental Impact Statement: Grizzly Bear*, 5–8. Diamond's quote is from Mary Hopkin, "Grizzly Bear Committee Supporters Outline Efforts," May 7, 1996. Fischer's quote is form Kenworthy, "Politics Imperils Uncommon Alliance's Plan to Find Grizzlies a Home," *Washington Post*, October 12, 1997. Salwasser's quote

is from "Reintroduction's the Recommendation," *Missoulian*, May 13, 1997. Johnson's quote is from Sherry Devlin, "Most Want More Area for Bears," *Missoulian*, October 3, 1997. The editorial quote is from D. F. Oliveria, "Participation Brings Measure of Control," *Spokane Spokesman-Review*, June 19, 1997. Church's quote is from Ben McNitt to Bear Group, June 30, 1997, HFP, ML. The outfitter's quote is from Hank Fischer, Press Advisory, July 24, 1997, HFP, ML.

36. Johnson's quote is from Sherry Devlin, "Most Want More Area for Bears," *Missoulian*, October 3, 1997. Schlickeisen's quote is from Rodger Schlickeisen to Michael Soule, March 7, 1997, HFP, ML.

37. The first quote is from Smith, "Groups Want Bears Protected," *Missoulian*, July 16, 1995. The middle two quotes are from Mark Matthews, "Strange Enemies Merging in Grizzly Reintroduction Talks," *Lewiston Tribune*, July 7, 1997. Craighead's quote is from "One Small Step by Humankind, One Giant Leap for the Great Bear," *Craighead Wildlife-Wildlands Institute* (Spring/Summer 1996): 2.

38. "Lemhi County Anthem," Minette Johnson Papers, Defenders of Wildlife Collection, Missoula MT.

39. The issue of rape was from Ken Miller, "Grizzlies Edge toward Idaho," *Idaho Statesman*, April 17, 1997. For the Nazi comparison, see Doug McClelland, "Grizzly Storm-Troopers' Poll Suspect," *Missoulian*, June 19, 1997. The quote from Salmon is from Minette Johnson to Hank Fischer and Tom France, October 9, 1997, HFP, ML. The issue of fear was from Minette Johnson to Hank Fischer, June 13, 1997, HFP, ML. Branch's quote is from Ric Branch, "Testimony for Oversight Hearing on Reintroduction of Grizzly Bear in the Public Domain National Forests," June 12, 1997, HFP, ML. For the IFBF quote, see Idaho Farm Bureau Federation, "Farm Bureau Opposes Grizzly Reintroduction," May 21, 1997, News Release, HFP, ML. The quote about land use is from Ken Miller, "Return of the Grizzly," *Idaho Statesman*, July 24, 1995. The final two quotes are from Minette Johnson to Hank Fischer and Tom France, October 9, 1997, HFP, ML; and Andrew M. Scutro, "Loaded for Bear: Salmon Rallies against Grizzlies," *Idaho Mountain Express*, October 15, 1997.

7. The Divided West

1. The statistics from the meetings are from U.S. Fish and Wildlife Service, *Final Environmental Impact Statement: Grizzly Bear*, 6-219.

2. The first quote is from U.S. Fish and Wildlife Service, *Final Environmental Impact Statement: Grizzly Bear*, 6–207. The second quote is from Charles Ellinger, "Grizzlies Next on Idaho's Import List," *Idaho Statesman*, February 11, 1995; and the following idea is from Andrew M. Scutro, "Loaded for Bear: Salmon Rallies against Grizzlies," *Idaho Mountain Express*, October 15, 1997. The final quote is from "Grizzly Bear Population Recovery."

3. Missoula's labor statistics are from, "Missoula County, 1990 Census of Population and Housing—Summary Set—Page #2: Selected Labor Force and Commuting Characteristics," accessed February 5, 2013, http://ceic.mt.gov/Demog/1990dec/stf/StfData/c032_2a3.htm. Missoula's education statistics are from, "Missoula City, 1990 Census of Population and Housing—Summary Set—Page #1: Selected Social Characteristics," accessed February 5, 2013, http://ceic.mt .gov/Demog/1990dec/stf/StfData/p108_1a3.htm.

4. This information is all from Robb, Riebsame, and Gosnell, eds., *Atlas of the New West*, 66, 113, 116, 119, 124, 127. For the effect of *A River Runs Through It*, see Brooks, *Bobos in Paradise*, 218–23. For perceptions of Missoula, see Chelsi Moy, "Missoula Reputation Tough to Overcome for Local Candidates," *Missoulian*, December 23, 2007.

5. The quote is from Short, "Growth and Development in Montana's Bitterroot Valley," 7, 12. The idea that Hamilton's population increase resulted from retirees is from Riebesame, *Atlas of the New West*, 57, 111. For information about the nature of Hamilton's population and its stance toward the government, see Richey, "Subdividing Eden," 93–100.

6. The first statistics are from "Ravalli County, 1990 Census of Population and House—Summary Set—Page #1: Selected Social Characteristic," accessed February 5, 2013, http://ceic.mt.gov/Demog /1990dec/STF/StfData/c041_1a3.htm; and "Ravalli County, 1990 Census of Population and Housing—Summary Set—Page #2: Selected Labor Force and Commuting Characteristics," accessed February 5, 2013, http://ceic.mt.gov/Demog/1990dec/STF/StfData /c041_2a3.htm. For the change in Hamilton's median age, see Swanson, *Growth and Change in the Bitterroot Valley*, 9. Hamilton's population statistics are from "Hamilton City, 1990 Census of Population and Housing—Summary Set—Page #1: Selected Social Characteris-

tics," accessed February 5, 2013, http://ceic.mt.gov/Demog
/1990dec/stf/StfData/p067_1a3.htm.

7. Robb, Riebsame, and Gosnell, eds., *Atlas of the New West*, 58; and
Hays, *Beauty, Health, and Permanence*, 32–35.

8. Brooks, *Bobos in Paradise*, 10, 18–21, 25–28.

9. For information about the rise of the new upper class, see Brooks,
Bobos in Paradise, 72, 203–4.

10. For the information about recreation habits, see Rothman, *Devil's
Bargains*, 169, 202–3, 226, 252–53.

11. Sarasohn, *Waiting for Lewis and Clark*, 8, 124.

12. The income statistics are from Travis, *New Geographies of the Ameri-
can West*, 15. Wyoming's statistics are from Paul Krza, "While the
New West Booms, Wyoming Mines, Drills . . . and Languishes," *High
Country News*, July 7, 1997. Montana and Idaho's statistics are from
Robb, Riebsame, and Gosnell, eds., *Atlas of the New West*, 108.

13. For information on Red Lodge, see Wiltsie and Wyckoff, "Reinventing Red Lodge," 125–50. For Jackson, see Culver, "From 'Last of the
Old West,'" 163–80. For Moab, see Amundson, "Yellowcake to Single Track," 151–62. Information about wolves is from Jones, "Way
Out West," 33–36.

14. Jon Christensen, "The Shotgun Wedding of Tourism and Public Lands," *High Country News*, December 23, 1996.

15. For the increase in second homes, see Rothman, *Devil's Bargains*,
233–36. Also see Brooks, *Bobos in Paradise*, 220–21. On population
statistics for Montana and Idaho, see Robb, Riebsame, and Gosnell,
eds., *Atlas of the New West*, 95–96.

16. Winkler et al., "Social Landscapes of the Inter-Mountain
West," 479–82, 498.

17. For reasons why conservatives moved to rural parts of the Intermountain West, see Sierra Crane-Murdoch, "Right-wing Migration," *High Country News*, May 13, 2013, 14–16. For information on
Hamilton, see Richey, "Subdividing Eden," 93–100.

18. Robb, Riebsame, and Gosnell, eds., *Atlas of the New West*, 106, 116,
119, 127, 130.

19. For the quotes, see Sarasohn, *Waiting for Lewis and Clark*, 136–37.
For Montana's economic statistics, see Swanson, *Growth and Change
in the Bitterroot Valley*, 11, 15, 16.

20. For more about the national culture wars of the 1990s, see Patterson, *Restless Giant*, 254–60; Flores, *Natural West*, 161; and Etulain and Malone, *American West*, 253–56.

21. Sarasohn, *Waiting for Lewis and Clark*, 127. Also see Rothman, *Devil's Bargains*, 10–12.

22. Information about Jackson is from Culver, "From 'Last of the Old West,'" 163–80. For information about rodeos and dude ranches, see Robb, Riebsame, and Gosnell, eds., *Atlas of the New West*, 121. Information about the Sundance catalog is from Nicholas, "1-800-SUNDANCE," 259–61.

23. Sarasohn, *Waiting for Lewis and Clark*, 136–37.

24. Ideas about authenticity or the contaminating influence of modernity and suspicion to outsiders is from Marston, "New Settlers in the Rural West," 79; and Charles Wilkinson, "Paradise Revised," 31. Also see Flores, foreword to *Imagining the Big Open*, viii.

25. For this quote, see Handley and Lewis, eds., *True West*, 1.

26. For more about commodification of the West, see Nicholas, "1-800-SUNDANCE," 265–68. For competing ideas of masculinity and authenticity, see Limerick, "Shadows of Heaven Itself," 155–57. For more ideas about authenticity, see Dan Flores, foreword to *Imagining the Big Open*, vii; and Nicholas, "1-800-SUNDANCE," 262.

27. The idea of New Westerners supporting the old model is from Travis, *New Geographies of the American West*, 23. The idea of Montana being anti-authority is from Brooks, *Bobos in Paradise*, 223.

28. For Turner's essay, see Frederick Jackson Turner, *Frontier in American History*, 1–38. For outlets of frontier anxiety, see Wrobel, *End of American Exceptionalism*, 43–68.

29. For the effect of the vast shift, see Flores, *Natural West*, 161. For Yellowstone's economy and economic transformation of the West at large, see Robb, Riebsame, and Gosnell, eds., *Atlas of the New West*, 107–8. Yellowstone's population statistics are from, "National Park Service Visitor Use Statistics: Yellowstone NP," accessed on February 19, 2013, https://irma.nps.gov/Stats/SSRSReports/Park%20specific%20Reports/Annual%20Park%20Visitation%20%28all%20years%29?Park=YELL. For the financial statistics for Bitterroot grizzlies, see U.S. Fish and Wildlife Service, *Final Environmental Impact Statement: Grizzly Bear*, xxiv.

30. The quote is from Stout, "Reintroduction of Grizzlies," 28. For the economic transformation, see Travis, *New Geographies of the American West*, 22–26.

8. Triumph and Collapse

1. Information about "Griz fever" is from Betsy Cohen, "Griz Trek Seems Like Old Times," *Missoulian*, December 18, 2001; and Rial Cummings, "Griz-Mania Uniquely Missoula," *Missoulian*, December 18, 2001. Rick Hill, "Oversight Hearing on Reintroduction of Grizzly Bear in the Public Domain National Forests," June 12, 1997, 159, HFP, ML.

2. Murray, *Great Bear*. For the resolution passed in Salmon, see Leslie Shumate, "Area Communities Form Anti-grizzly Coalition," *Salmon (ID) Recorder-Herald*, September 2, 1999.

3. U.S. Fish and Wildlife Service, *Final Environmental Impact Statement: Grizzly Bear*, 6-215-2-220.

4. For support for Alternative 4—the conservation biology alternative—see Rhonda Swaney to Chris Servheen, October 23, 1997, HFP, ML; Nez Perce Tribal Executive Committee to Chris Servheen, November 25, 1997, HFP, ML; and World Wildlife Fund to Chris Servheen, October 28, 1997, HFP, ML. For more objections, see Brian Peck and Louisa Willcox to Chris Servheen, September 30, 1997, HFP, ML.

5. The original letter is from "Canadian Environmental Organizations Oppose Removal of Grizzly Bears from British Columbia for US Recovery Efforts," News Release, December 5, 1996, HFP, ML. Reasons for not wanting bears to come from the NCDE and GYE are found in Rhonda Swaney to Chris Servheen, October 23, 1997; and World Wildlife Fund to Chris Servheen, October 28, 1997, HFP, ML. For more objections, see Brian Peck and Louisa Willcox to Chris Servheen, September 30, 1997, HFP, ML.

6. The idea of taking bears from Alaska is from Sterling Miller to Tom France, December 8, 1997, HFP, ML. For more in-depth discussion, see Tom France, Sterling Miller, and Hank Fischer to Louisa Willcox, March 2, 1998, HFP, ML.

7. Rodger Schlickeisen to Bill Meadows, July 3, 1997, HFP, ML.

8. The first part of the paragraph is from Louisa Willcox to Tom France and Sterling Miller, December 24, 1998, HFP, ML. Nez Perce Tribal Executive Committee to Chris Servheen, November 25, 1997, HFP ML.

9. Conservative critiques of citizen management committee are from Jerry Miller, "Grizzly Return Seen as 'Trojan Horse' for Bruce Babbitt," *Messenger Index*, July 2, 1997. Complaints from environmentalists about the citizen management committee are from William H. Meadows to Mark Van Putten and Rodger Schlickeisen, June 27, 1997, HFP, ML. The committee's legality is from Hall, "Subdelegation of Authority under the Endangered Species Act," 83, 85.

10. The majority of the paragraph is taken from Sherry Devlin, "Grizzlies in Our Midst," *Missoulian*, April 30, 1998. Servheen's final quote is from Beth Wohlberg, "Caught in a Bear Trap," *Missoula Independent*, March 11, 1999. The last fact is from Timothy Floyd, "Stick to the Bear Facts," *Post Register*, August 13, 1997.

11. For information about meetings, see Ken Dey, "Learning to Live in Bear Country," *Ravalli Republic*, July 30, 1997. Also see interview with Claire Kelly, October 23, 2013, in possession of the author.

12. For David Quammen's argument and quote, see Quammen, *Monsters of God*, 6. The first and last quotes is from Tim Woodward, "It Doesn't Take Many Griz to Be More than Enough," *Idaho Statesman*, July 13, 1997. The second quote is from Tom Kenworthy, "Politics Imperils Uncommon Alliance's Plan to Find Grizzlies a Home," *Washington Post*, October 12, 1997. Chenoweth's quote is from Guy Gugliotta, "Lawmaker Finds Plan Hard to Swallow," *Washington Post*, September 9, 1997.

13. Racicot's objections are from Marc Racicot to Jamie Clark and Ralph Morgenweck, October 7, 1997, HFP, ML. For a response, see Jamie Clark to Marc Racicot, November 21, 1997, HFP, ML.

14. Craig's quote is from Brandon Loomis, "Craig Tries to Block Funds for Grizzly Reintroduction," *Lewiston Morning Tribune*, July 25, 1997. For information on the bill and cuts, see Defenders of Wildlife, "President Signs Appropriations Bill with Riders Harmful to Public Lands & Wildlife," News Release, November 14, 1997, HFP, ML; and "Grizzly Recovery Program Loses Half of Funding," *Missoulian*, January 12, 1998. Information about Bitterroot funding is from Jamie Clark to Marc Racicot, November 12, 1997, HFP, ML.

15. ROOTS coalition to Chris Servheen, October 29, 1997, HFP, ML. Information about the sanitation program is from, U.S. Fish and Wildlife Service, *Grizzly Bear Recovery in the Bitterroot Ecosystem*, 8;

and Hank Fischer to Rodger Schlickeisen and Bob Ferris, January 29, 1998, HFP, ML.

16. "Feds Failed to Consult Idaho on Grizzlies? Gadzooks," *Lewiston Morning Tribune*, July 18, 1997.

17. Information about Baucus is from Hank Fischer to Jim Riley, November 10, 1997, HFP, ML. Information about the governorship is from Hank Fischer to Rodger Schlickeisen and Bob Ferris, January 29, 1998, HFP, ML.

18. Information about Mealey is from "New Wildlife Chief to Walk Tightrope," *Spokane Spokesman-Review*, February 3, 1997. His quote is from McNamee, "Breath-holding," 77. His threat to block reintroduction is from Steve Bard, "F&G Chief: No Grizzlies for Idaho," *Idaho Statesman*, August 23, 1997.

19. For Defenders of Wildlife's strategizing, see Hank Fischer to Rodger Schlickeisen and Bob Ferris, January 29, 1998, HFP, ML. Information about planned release is from and email from Chris Servheen to Jill Parker, February 19, 1998, HFP, ML. For more information, see Tom France to Greg Schildwachter, Bill Wall, Rodger Schlickeisen, and Bob Ferris, March 17, 1998, HFP, ML. Information about Kempthorne is from Hank Fischer to Max Baucus, June 9, 1998, HFP, ML.

20. Jim Riley, Tom France, Hank Fischer, and Bill Mulligan to Ralph Regula and Sidney Yates, May 21, 1998, HFP, ML.

21. Rodger Schlickeisen, Bill Wall, Greg Schildwachter, and Tom France to Jamie Clark, April 21, 1998; Hank Fischer to Jamie Clark, June 4, 1998, HFP, ML. Also, see the interview with Hank Fischer, March 2, 2012, in possession of the author.

22. Hank Fischer to Max Baucus, June 9, 1998; and Hank Fischer to Scott Conroy, June 8, 1998, HFP, ML.

23. For information on Rick Hill and Michael Crapo, see Hank Fischer to Tom France, Mary Beth Beetham, Sara Barth, Minette Johnson, and Mike Senatore, April 21, 1998, HFP, ML. Information about Burns is from Fischer to Max Baucus, June 9, 1998; and Greg Schildwachter to Hank Fischer and Sterling Miller (email), June 6, 1998, HFP, ML. Information about the rider is from Defenders of Wildlife, "Congress Ignores White House Veto Threat, Loads Bill with 'Worst-Ever' Anti-Environmental Riders," News Release, June 24, 1998, HFP, ML.

The White House's opposition is from Rodger Schlickeisen to Hank Fischer (email), June 16, 1998, HFP, ML.

24. For Custer County's actions, see "Commissioners Pass Ordinance to Bar Grizzlies," *Spokane Spokesman-Review*, January 20, 1998. Information about Walters's bill is from Beth Wohlberg, "Caught in a Bear Trap," *Missoula Independent*, March 11, 1999.

25. Sherry Devlin, "GOP Senators Say Griz Study Is Incomplete," *Missoulian*, February 27, 1999; and Sherry Devlin, "GOP-requested Study Shows There's Room for the Great Bear," *Missoulian*, February 19, 1999.

26. The call for delisting is from Conrad Burns, Larry Craig, Michael Crapo, and Craig Thomas to Bruce Babbitt, March 2, 1999, HFP, ML. Also see Devlin, "GOP-requested Study Shows There's Room for the Great Bear," *Missoulian*, February 19, 1999; and Tom France and Sterling Miller, "Op Ed: Grizzly Bear Habitat Suitability in the Bitterroots," April 5, 1999, HFP, ML. The last fact is from Michael Milstein, "Thomas and Burns Join Opposition to Adding Grizzlies to Selway," *Gazette Wyoming Bureau*, March 4, 1999.

27. Attempt to gain Babbitt's support is from Hank Fischer to Bob Ferris and Minette Johnson, April 13, 1999; and Hank Fischer to Rodger Schlickeisen (email), September 9, 1999, HRF, ML. For Clark's delay and Clinton's stance, see David Knibb, *Grizzly Wars*, 153, 150. Influence from Kempthorne is from the interview with Hank Fischer, March 2, 2012, in possession of the author. Babbitt's unwillingness to push Clark is from Mark Van Putten to Tom France (email), December 22, 1999, HFP, ML.

28. Kempthorne's position and quote is from Alliance for the Wild Rockies, "Idaho Governor on Record Against Bitterroot Grizzly Reintroduction," News Release, September 30, 1999, HRP, ML.

29. For more on Andrus, see "Gov. Andrus Cool to Grizzly Bear Recovery Plans," *Lewiston Morning Tribune*, May 16, 1993; Hays, *Beauty, Health, and Permanence*, 50; and Rocky Barker, "Cecil Andrus Knew How to Take a Stand," *High Country News*, February 20, 1995.

30. Election information comes from "2000 Presidential Election: Popular Vote Totals," U.S. National Archives and Records Administration, accessed June 15, 2014, http://www.archives.gov/federal-register/electoral-college/2000/popular_vote.html; and "U.S. Presidential Election Results," Dave Leip's Atlas of U.S. Presidential

Elections, accessed June 15, 2014, http://uselectionatlas.org
/RESULTS/.

31. For information about Crapo, see Conrad Burns, Larry Craig,
Michael Crapo, and Craig Thomas to Bruce Babbitt, March 2, 1999,
HFP, ML. Information about the meeting in Salmon is from Leslie
Shumate, "Area Communities Form Anti-grizzly Coalition," *Salmon
(ID) Recorder-Herald*, September 2, 1999.

32. Sherry Devlin, "Wolf Ruling Won't Apply to Grizzlies," *Missoulian*,
March 5, 1998. For the announcement of the search, see Sherry Devlin, "Are Grizzly Bears Already in Bitterroot," *Missoulian*, November
2, 1999. Information about the actual search is from "The Bears Out
There," *Missoula Independent*, June 21, 2001.

33. Servheen's first quote is from Stout, "Reintroduction of Grizzlies,"
4. His second quote is from Sherry Devlin, "What Grizzly Bears?"
Missoulian, November 3, 1999. Information about the search and
Campbell's quote is from Steven Allison-Bunnell, "The Griz Files,"
Missoula Independent, December 2, 1999. The final quote is from
Sherry Devlin, "Rules of the Hunt," *Missoulian*, November 18, 1999.

34. Steven Allison-Bunnell, "The Griz Files," *Missoula Independent*,
December 2, 1999.

35. Fischer's comment is from Hank Fischer to Phil Church, Bill Mulligan, Jim Riley, and Tom France, November 3, 1999, HFP, ML. Also,
see "The Bears Out There," *Missoula Independent*, June 21, 2001.

36. U.S. Fish and Wildlife Service, *Final Environmental Impact
Statement: Grizzly Bear*, xv, xxxiv.

37. The first quote is from Greg Nelson, "Support Governor Batt's
Grizzly Stance," May 23, 1997. The pseudospirituality comment is
from Mike Costello, "Introduce the Grizzly and We'll All Work for
Gore," *Lewiston Tribune*, January 25, 1997. The second-to-last quote
is from D. F. Oliveria, "Grizzlies Make Lethal Neighbors," *Spokane
Spokesman-Review*, March 24, 2000. The last quote is from Velado,
"Grizzly Bear Reintroduction to the Bitterroot Ecosystem," 82.

38. Bader's quote is from Walter Kuhlman, Memorandum: Legal and
Factual Issues related to Grizzlies in the Northern Rockies, April 30,
1996, HFP, ML. Peck's quote is from Carlotta Grandstaff, "God, Science, and Grizzlies," *Bitterroot Star*, October 8, 1997. Fischer's quote
is from Guy Gugliotta, "Lawmaker Finds Plan Hard to Swallow,"
Washington Post, September 9, 1997. The final quote is from Bill

Loftus, "Grizzly Recovery Favored at Hearing," *Lewiston Morning Tribune*, October 3, 1997. The poll is from Idaho Consulting International, Idaho Voter Opinion Poll, February 6, 1997, HFP, ML.

39. Hank Fischer to Tom France (email), April 24, 2000, and Hank Fischer to Tom France (email), May 14, 2000, HFP, ML.

40. For the strategy, see Hank Fischer to Tom France (email), May 14, 2000, HFP, ML. For more about the funding, see Sterling Miller to Mark Boyce (email), June 30, 2000; and Sterling Miller to Mark Boyce (email), July 5, 2000, HFP, ML.

41. For Racicot's support, see Mark Van Putten to Tom France (email), December 22, 1999, HFP, ML. Kempthorne's two quotes are from Dan Hansen, "Kempthorne to Grizzlies: Keep out!" *Spokane Spokesman-Review*, November 17, 2000; and Sherry Devlin, "Grizzlies OK'd for Bitterroot," *Missoulian*, November 17, 2000. The quip at Kempthorne is from Dick Dorworth, "Economics True Issue in Opposing Grizzly Reintroduction," *Idaho Mountain Express*, November 29, 2000. Information about Kempthorne's lawsuit is from Dirk Kempthorne, "Kempthorne Says Idaho Will Go to Court to Stop Feds Grizzly Bear Plan," News Release, November 16, 2000.

42. For the USFWS's appeal to Kempthorne, see Ralph O. Morgenweck to Dirk Kempthorne, December 13, 2000, HFP, ML. For the decision to hire a law firm, see Melanie Carroll, "Group Will Hire Lawyers to Fight Grizzly Introduction," *Missoulian*, December 13, 2000. For the merits of the lawsuit, see Knibb, *Grizzly Wars*, 168; and Sherry Devlin, "Groups Fight to Keep Bear Reintroduction," *Missoulian*, March 14, 2001.

43. Information on the Bush administration is from "Will Bush's Principles Bear Up?" *Missoulian*, May 7, 2001. Norton's quote and Kempthorne's influence are from Sherry Devlin, "Reintroduction under Scrutiny," *Missoulian*, April 26, 2001.

44. Norton's decision and France's quotes are from Associated Press, "Norton Proposes Scrapping Griz Plan," *Missoulian*, June 21, 2001; and Sherry Devlin, "Norton Snubs Grizzlies," *High Country News*, July 30, 2001. Miller's quote is from Sterling Miller, Guest Column, *Missoulian*, November 5, 2001.

45. U.S. Fish and Wildlife Service, *Summary of Public Comments on the Notice of Intent to Re-Evaluate the Record of Decision for Grizzly Bear*

Recovery in the Bitterroot Ecosystem, October 2001, 13. For the Sierra Club's position, see Sierra Club, "Sierra Club Condemns Secretary of Interior's Attack on Grizzly Bear Reintroduction Program," June 20, 2001, HFP, ML. France's quote is from Sherry Devlin, "Public Sounds Off on Grizzly Proposal," *Missoulian,* October 23, 2001.

46. For the relationship between the election and bears, see Knibb, *Grizzly Wars,* 164, and Hansen, "Kempthorne to Grizzlies: Keep Out!" *Spokane Spokesman-Review,* November 17, 2000.

Conclusion

1. Perry Backus, "Grizzly Shot in Selway-Bitterroot," *Missoulian,* September 12, 2007.

2. For information about the bear's travels, see Michael Jamison, "Bitterroot Griz Logged at Least 140 Miles," *Missoulian,* October 4, 2007. For the search, see "Bitterroot Mountains: 51 Cameras Can't Catch Grizzly Bears," *Missoulian,* November 21, 2008.

3. Information about the grizzly killed is from Michael Jamison, "Bitterroot Griz Logged at Least 140 Miles," *Missoulian,* October 4, 2007. Information about the EIS is from the interview with Chris Servheen, March 26, 2012, in possession of the author.

4. For headwaters' study, see Jeff Welsch, "Thank You, James Watt, for All You Did for Great Yellowstone," *High Country News,* August 1, 2013. For the 2013 survey, see Renata Birkenbuel, "Outdoors Big Draw to Small Biz: 70% Say State's Enticing Environment Is a Factor," *Montana Standard,* September 11, 2013.

5. Travis, *New Geographies of the American West,* 15, 23, 160.

6. Leslie Shumate, "Area Communities Form Anti-Grizzly Coalition," *Salmon (ID)Recorder-Herald,* September 2, 1999.

7. For information about the Blackfoot Challenge, see Keila Szpaller, "Interior Secretary Salazar Touts Conservation Partnerships in Ovando," *Missoulian,* July 16, 2011. Information about caribou is from Kofinas, "Caribou Hunters and Researchers at the Co-management Interface," 190. For information about the Charles C. Deam Wilderness, see Slover, "Music of Opinions," 18–23.

8. See Dan Flores's chapter, "Dreams and Beasts," in his book *The Natural West,* 74, 81.

9. Nash, *Rights of Nature,* 8.

Bibliography

Archives

Boise State University Archives and Special Collections
    Cecil Andrus Collection
    Larry LaRocco Papers
University of Montana, Maureen and Mike Mansfield Library, Archives
    and Special Collections
    Hank Fischer Papers
    Doris and Kelsey C. Milner Papers
    Pat Williams Papers

Published Sources

Amundson, Michael A. "Yellowcake to Single Track: Culture, Community, and Identity in Moab Utah." In *Imagining the Big Open: Nature, Identity, and Play in the West*, edited by Liza Nicholas, Elaine M. Bapis, and Thomas J. Harvey, 151–62. Salt Lake City: University of Utah Press, 2003.

Askins, Renee. "Releasing Wolves from Symbolism." *Harper's Magazine* 290, no. 1739 (April 1995).

Bader, Mike, Michael Garrity, and Timothy Bechtold. *The Conservation Biology Alternative for Grizzly Bear Population Restoration in the Greater Salmon-Selway Region of Central Idaho and Western Montana*. Missoula MT: Alliance for the Wild Rockies, January 1996.

Barnes, Fred. "Revenge of the Squares." *New Republic*, March 13, 1995.

Bean, Michael J. "Looking Back over the First Fifteen Years." In *Balancing on the Brink of Extinction: The Endangered Species Act and Lessons for the Future*, edited by Kathryn A. Kohm, 37–42. Washington DC: Island Press, 1991.

Begley, Sharon. "The Return of the Native." *Newsweek* 125, no. 4 (January 23, 1995).

Biel, Alice Wondrak. *Do (Not) Feed the Bears: The Fitful History of Wildlife and Tourists in Yellowstone*. Lawrence: University Press of Kansas, 2006.

Bixby, Kevin. "Predator Conservation." In *Balancing on the Brink of Extinction: The Endangered Species Act and Lessons for the Future*, edited by Kathryn A. Kohm, 199–213. Washington DC: Island Press, 1991.

Brooks, David. *Bobos in Paradise: The New Upper Class and How They Got There*. New York: Simon & Schuster, 2000.

Cauble, Christopher. "Return of the Native: The Parks Service Calls for Reintroduction of Yellowstone's Missing Predator . . . the Wolf." *National Parks*, July/August 1986.

Cawley, R. McGreggor. *Federal Land, Western Anger: The Sagebrush Rebellion and Environmental Politics*. Lawrence: University Press of Kansas, 1993.

Cestero, Barb. *Beyond the Hundredth Meeting: A Field Guild to Collaborative Conservation on the West's Public Lands*. Tucson AZ: Sonoran Institute, July 1999.

Chris, Cynthia. *Watching Wildlife*. Minneapolis: University of Minnesota Press, 2006.

Clark, Ella E. *Indian Legends of the Northern Rockies*. Norman: University of Oklahoma Press, 1966.

Commoner, Barry. "The Broader Context of the Environmental Movement." In *The American Environment: Readings in the History of Conservation*, edited by Roderick Nash, 238–46. 2nd ed. Reading MA: Addison-Wesley, 1976.

Conley, Alexander, and Margaret A. Moote. "Evaluating Collaborative Natural Resource Management." *Society and Natural Resources* 16, no. 5 (2003): 371–86.

Culver, Lawrence. "From 'Last of the Old West' to First of the New West: Tourism and Transformation in Jackson Hole, Wyoming." In *Imagining the Big Open: Nature, Identity, and Play in the West*, edited by Liza Nicholas, Elaine M. Bapis, and Thomas J. Harvey, 163–80. Salt Lake City: University of Utah Press, 2003.

Dant, Sara. "Making Wilderness Work: Frank Church and the American Wilderness Movement." *Pacific Historical Review* 77, no. 2 (May 2008).

Egan, Timothy. *Lasso the Wind: Away to the New West*. New York: Knopf, 1998.

Etulain, Richard W., and Michael P. Malone. *The American West: A Modern History, 1900 to the Present*. 2nd ed. Lincoln: University of Nebraska Press, 1989.

Fischer, Hank. "Bears and the Bitteroot." *Defenders Magazine*, Winter 1996/97.

———. "Moving Past the Polarization: Wolves, Grizzly Bears, and the Endangered Species Act." Paper presented at the Wallace Stegner Center Symposium, Salt Lake City, April 13, 1996.

———. *Wolf Wars: The Remarkable Inside Story of the Restoration of Wolves to Yellowstone*. Helena MT: Falcon, 1995.

Fischer, Hank, and Mike Roy. "New Approaches to Citizen Participation in Endangered Species Management: Recovery in the Bitterroot Ecosystem." *Ursus* 10 (1998).

Flores, Dan. Foreword to *Imagining the Big Open: Nature, Identity, and Play in the West*, edited by Liza Nicholas, Elaine M. Bapis, and Thomas J. Harvey, vii–ix. Salt Lake City: University of Utah Press, 2003.

———. *The Natural West: Environmental History in the Great Plains and Rocky Mountains*. Norman: University of Oklahoma Press, 2001.

"Grizzly Bear Population Recovery: The Vital Link with Roadless Areas of the Greater Salmon-Selway Ecosystem." *Big Wild Action Report* (February 1996).

Guynn, Dwight E., and Marion K. Landry. "A Case Study of Citizen Participation as a Success Model for Innovative Solutions for Natural Resource Problems." *Wildlife Society Bulletin* 25, no. 2 (Summer 1997).

Hall, Brenda Lindlief. "Subdelegation of Authority under the Endangered Species Act: Secretarial Authority to Subdelegate His Duties to a Citizens Management Committee as Proposed for the Selway-Bitterroot Wilderness Grizzly Bear Reintroduction." *Public Land & Resources Law Review* 20, no. 1 (1999).

Handley, William R., and Nathaniel Lewis, eds. *True West: Authenticity and the American West*. Lincoln: University of Nebraska Press, 2003.

Haynes, Bessie Doak, and Edgar Haynes, eds. *The Grizzly Bear: Portraits from Life*. Norman: University of Oklahoma Press, 1966.

Hays, Samuel P. *Beauty, Health, and Permanence: Environmental Politics in the United States, 1955–1985*. Cambridge: Cambridge University Press, 1987.

———. "Conservation as Efficiency." In *The American Environment: Readings in the History of Conservation*, edited by Roderick Nash, 82–84. 2nd ed. Reading MA: Addison-Wesley, 1976.

Hibbard, Michael, and Jeremy Madsen. "Environmental Resistance to Place-Based Collaboration in the U.S. West." *Society and Natural Resources* 16, no. 8 (2003): 703–18.

Holthaus, Gary H., Charles F. Wilkinson, Patricia Nelson Limerick, ed. *Society to Match the Scenery: Personal Visions of the Future of the American West*. Boulder: University Press of Colorado, 1991.

Houck, Oliver A. "Reflections on the Endangered Species Act." *Environmental Law* 25 (Summer 1995).

Interagency Grizzly Bear Committee. *Looking Back: Twenty-five Years of the Interagency Grizzly Bear Committee*. Missoula MT: IGBC, 2008.

Irving, Washington. "Adventures of William Cannon and John Day with Grizzly Bears." In *The Grizzly Bear: Portraits from Life*, edited by Bessie Doak Haynes and Edgar Haynes, 25–29. Norman: University of Oklahoma Press, 1966.

———. *Astoria; or, Anecdotes of an Enterprise Beyond the Rocky Mountains*. Aurora CO: Bibliographical Center for Research, 2009. First published 1836.

Jones, Karen. "Way Out West . . . Ghost Towns, Gray Wolves, Territorial Prisons and More!" In *Imagining the Big Open: Nature, Identity, and Play in the West*, edited by Liza Nicholas, Elaine M. Bapis, and Thomas J. Harvey, 27–44. Salt Lake City: University of Utah Press, 2003.

Kagan, Robert A. "Political and Legal Obstacles to Collaborative Ecosystem Planning." *Ecology Law Quarterly* 24 (1997): 871–75.

Kellert, Stephen R. *Public Attitudes Toward Critical Wildlife and Natural Habitat Issues*. Washington DC: U.S. Department of the Interior, Fish and Wildlife Service, 1979.

———. *The Value of Life: Biological Diversity and Human Society.* Washington DC: Island Press, 1996.

Kline, Phillip. "Grizzly Bear Blues: A Case Study of the Endangered Species Act's Delisting Process and Recovery Plan Requirements." *Environmental Law* 31, no. 371 (Spring 2001).

Kloppenberg, James T. "An Old Name for Some New Ways of Thinking?" *Journal of American History* 83, no. 1 (June 1996).

Knibb, David. *Grizzly Wars: The Public Fight over the Great Bear.* Spokane: Eastern Washington University Press, 2008.

Kofinas, Gary P. "Caribou Hunters and Researchers at the Co-management Interface: Emergent Dilemmas and the Dynamics of Legitimacy in Power Sharing." *Anthropologica* 47, no. 2 (2005).

Kohm, Kathryn A. "The Act's History and Framework." In *Balancing on the Brink of Extinction: The Endangered Species Act and Lessons for the Future,* edited by Kathryn A. Kohm. Washington DC: Island Press, 1991.

———, ed. *Balancing on the Brink of Extinction: The Endangered Species Act and Lessons for the Future.* Washington DC: Island Press, 1991.

Kunkel, Kyran E., Wendy Clark, and Gregg Servheen. *A Remote Survey for Grizzly Bears in Low Human Use Areas of the Bitterroot Grizzly Bear Evaluation Area.* Boise: Idaho Department of Fish and Game, 1991.

Lapinski, Mike. *Grizzlies and Grizzled Old Men: A Tribute to Those who Fought to Save the Great Bear.* Helena MT: Falcon Press Publishing, 2006.

Leopold, Aldo. *A Sand County Almanac with Essays on Conservation from Round River.* New York: Ballantine Books, 1966.

Limerick, Patricia Nelson. "The Shadows of Heaven Itself." In *Atlas of the New West: Portrait of a Changing Region,* edited by James Robb, William E. Riebsame, and Hannah Gosnell, 151–78. New York: W. W. Norton, 1997.

Lummis, Charles Fletcher. "Begging the Bear's Pardon." In *The Grizzly Bear: Portraits from Life,* edited by Bessie Doak Haynes and Edgar Haynes, 333–34. Norman: University of Oklahoma Press, 1966.

Mann, Charles C., and Mark L. Plummer. "Is Endangered Species Act in Danger?" *Science* 267 (March 3, 1995): 1256–58.

———. *Noah's Choice: The Future of Endangered Species.* New York: Knopf, 1996.

Marsh, James B. *Four Years in the Rockies; or, The Adventures of Isaac P. Rose.* New Castle PA: W. B. Thomas, 1884.

Marshall, Robert. "Wilderness." In *The American Environment: Readings in the History of Conservation*, edited by Roderick Nash, 121–26. 2nd ed. Reading MA: Addison-Wesley, 1976.

Marston, Betsy. "New Settlers in the Rural West." In *Society to Match the Scenery: Personal Visions of the Future of the American West*, edited by Gary Holthaus, Charles Wilkinson, and Patricia Limerick, 78–80. Boulder: University Press of Colorado, 1991.

McNamee, Thomas. "Breath-holding in Grizzly Country." *Audubon* 84, no. 6 (November 1982): 77.

———. *The Return of the Wolf to Yellowstone*. New York: Henry Holt, 1997.

McNaught, David A. "Wolves in Yellowstone? Park Visitors Respond." *Wildlife Society Bulletin* 15, no. 4 (Winter 1987).

Melquist, Wayne E. *A Preliminary Survey to Determine the Status of Grizzly Bears (Ursus Arctos Horribilis) in the Clearwater National Forest of Idaho*. Moscow: University of Idaho, 1985.

Mighetto, Lisa. *Wild Animals and American Environmental Ethics*. Tucson: University of Arizona Press, 1991.

Miller, Joaquin. "A Grizzly's Sly Little Joke." In *The Grizzly Bear: Portraits from Life*, edited by Bessie Doak Haynes and Edgar Haynes, 331–32. Norman: University of Oklahoma Press, 1966.

Mills, Enos A. *The Grizzly: Our Greatest Wild Animal*. New York: Houghton Mifflin, 1919.

———. "Trailing without a Gun." In *The Grizzly Bear: Portraits from Life*, edited by Bessie Doak Haynes and Edgar Haynes. Norman: University of Oklahoma Press, 1966.

Moore, Bud. "Last of the Bitterroot Grizzly." *Montana Magazine*, November/December 1984.

———. *The Lochsa Story: Land Ethics in the Bitterroot Mountains*. Missoula MT: Mountain Press Publishing, 1996.

Moote, Margaret A., and Mitchel P. McClaren. "Viewpoint: Implications of Participatory Democracy for Public Land Planning." *Journal of Range Management* 50, no. 5 (September 1997): 473–81.

Murray, John A., ed. *The Great Bear: Contemporary Writings on the Grizzly*. Anchorage: Alaska Northwest Books, 1992.

Nash, Roderick Frazier, ed. *The American Environment: Readings in the History of Conservation*. 2nd ed. Reading MA: Addison-Wesley, 1976.

———. *The Rights of Nature: A History of Environmental Ethics*. Madison: University of Wisconsin Press, 1989.

Nicholas, Liza. "1-800-SUNDANCE: Identity, Nature, and Play in the West." In *Imagining the Big Open: Nature, Identity, and Play in the West*, edited by Liza Nicholas, Elaine M. Bapis, and Thomas J. Harvey, 259–71. Salt Lake City: University of Utah Press, 2003.

Nicholas, Liza, Elaine M. Bapis, and Thomas J. Harvey. eds. *Imagining the Big Open: Nature, Identity, and Play in the West*. Salt Lake City: University of Utah Press, 2003.

Nixon, Richard. "51-Special Message to the Congress Outlining the 1972 Environmental Program." February 8, 1972. The American Presidency Project. Accessed on March 23, 2013. http://www.presidency.ucsb.edu/ws/index.php?pid=3731.

Nixon, Richard, and the Council on Environmental Quality. "Environmental Priorities for the 1970s." In *The American Environment: Readings in the History of Conservation*, edited by Roderick Nash, 247–64. 2nd ed. Reading MA: Addison-Wesley, 1976.

O'Connell, Kim A. "Wise Use Groups Lobby Against Species Act." *National Parks*, September/October 1995.

O'Toole, Randall, and Karyn Moskowitz. "Beyond the 100th Paradigm." *Different Drummer* (Summer 1996): 3–7.

"One Small Step by Humankind, One Giant Leap for the Great Bear." *Craighead Wildlife-Wildlands Institute*, Spring/Summer, 1996.

Orloff, Neil. *The Environmental Impact Statement Process: A Guide to Citizen Action*. Washington DC: Information Resources Press, 1978.

Patterson, James T. *Restless Giant: The United States from Watergate to Bush v. Gore*. New York: Oxford University Press, 2005.

Peacock, Doug, and Andrea Peacock. *The Essential Grizzly: The Mingled Fates of Men and Bears*. Guilford CT: Lyons Press, 2006.

Petersen, M. Nils, Markus J. Petersen, and Taria Rai Peterson. "Conservation and the Myth of Consensus." *Conservation Biology* 19, no. 3 (June 2005).

Petersen, Shannon. *Acting for Endangered Species: The Statutory Ark*. Lawrence: University of Kansas Press, 2002.

Pinchot, Gifford. "Ends and Means." In *The American Environment: Readings in the History of Conservation*, edited by Roderick Nash, 58–63. 2nd ed. Reading MA: Addison-Wesley, 1976.

"Public Opinions and Attitudes toward Reintroducing Grizzly Bears to the Selway Bitterroot Wilderness Area of Idaho and Montana." *Responsive Management* (April 1997).

"Public Knowledge, Opinion, and Attitudes toward Endangered Species Reintroduction: Two Case Studies." *Responsive Management* (Winter 1996).

Quammen, David. *Monsters of God: The Man-Eating Predator in the Jungles of History and the Mind*. New York: W. W. Norton, 2003.

Richey, E. Duke. "Subdividing Eden: Land Use and Change in the Bitterroot Valley, 1930–1998." Master's thesis, University of Montana, May 1998.

Robb, James, William E. Riebsame, and Hannah Gosnell, eds. *Atlas of the New West: Portrait of a Changing Region*. New York: W. W. Norton, 1997.

Robinson, Michael. *Predatory Bureaucracy: The Extermination of Wolves and the Transformation of the West*. Boulder: University Press of Colorado, 2005.

Rohlf, Daniel J. *The Endangered Species Act: A Guide to Its Protections and Implementation*. Stanford: Stanford Environmental Law Society, 1989.

Roosevelt, Theodore. "Hunting in the West." In *The Grizzly Bear: Portraits from Life*, edited by Bessie Doak Haynes and Edgar Haynes. Norman: University of Oklahoma Press, 1966.

Rothman, Hal K. *Devil's Bargains: Tourism in the Twentieth-Century American West*. Lawrence: University Press of Kansas, 1998.

———. *The Greening of a Nation? Environmentalism in the United States Since 1945*. Fort Worth: Harcourt Brace College Publishers, 1998.

Ruxton, George Frederick Augustus. "The Saga of Hugh Glass." In *The Grizzly Bear: Portraits from Life*, edited by Bessie Doak Haynes and Edgar Haynes, 50–56. Norman: University of Oklahoma Press, 1966.

Sarasohn, David. *Waiting for Lewis and Clark: The Bicentennial and the Changing West*. Portland: Oregon Historical Society Press, 2005.

Schullery, Paul. *Lewis and Clark Among the Grizzlies: Legend and Legacy in theAmerican West*. Helena MT: Falcon Press, 2002.

———. *The Yellowstone Wolf: A Guide and Sourcebook*. Norman OK: Red River Books, 1996.

Servheen, Christopher. *Grizzly Bear Recovery Plan*. Missoula MT: U.S. Fish and Wildlife Service, 1993.

———. *Grizzly Bear 5-year Review: Summary and Evaluation*. Missoula MT: U.S. Fish and Wildlife Grizzly Bear Recovery Office, 2011.

Short, D. C. "Growth and Development in Montana's Bitterroot Valley: The Valley Is Booming—But Is It a Bust for the Locals?" Master's thesis, University of Montana, 1988.

Simpson, Alan. "Wyoming Wolves? 'No!'" *Wyoming Wildlife Magazine*, March 1988.

Slover, Bruce L. "A Music of Opinions: Collaborative Planning for the Charles C. Deam Wilderness." *Journal of Forestry* 94, no. 5 (May 1996): 18–23.

Space, Ralph S. *The Clearwater Story: A History of the Clearwater National Forest*. Missoula: U.S. Department of Agriculture, Forest Service, 1964.

———. *Lolo Trail: A History of Events Connected with the Lolo Trail*. Lewiston: Printcraft Printing, 1970.

Stegner, Wallace. "The Meaning of Wilderness in American Civilization." In *The American Environment: Readings in the History of Conservation*, edited by Roderick Nash, 192–96. 2nd ed. Reading MA: Addison-Wesley, 1976.

Stout, Ray R. "The Reintroduction of Grizzlies to the Selway-Bitterroot Wilderness: A Boon or a Burden?" Master's thesis, University of Montana, May 1996.

Sutter, Paul S. *Driven Wild: How the Fight against Automobiles Launched the Modern Wilderness Movement*. Seattle: University of Washington Press, 2002.

Swanson, Larry. *Growth and Change in the Bitterroot Valley and Implications for Area Agriculture and Ag Lands*. Missoula MT: O'Connor Center for the Rocky Mountain West, University of Montana, April 2006.

Travis, William R. *New Geographies of the American West*. Washington DC: Island Press, 2007.

Turner, Frederick Jackson. *The Frontier in American History*. New York: Holt, Rinehart, and Winston, 1947. First published 1909.

Udall, Stewart L. "Prospects for the Land." In *The American Environment: Readings in the History of Conservation*, edited by Roderick Nash, 210–22. 2nd ed. Reading MA: Addison-Wesley, 1976.

U.S. Fish and Wildlife Service. *Final Environmental Impact Statement: Grizzly Bear Recovery in the Bitterroot Ecosystem*. Washington DC: U.S. Government Printing Office, 2000.

———. *Final Environmental Impact Statement: The Reintroduction of Gray Wolves to Yellowstone National Park and Central Idaho*. Washington DC: U.S. Government Printing Office, 1994.

———. *Final Rule on Establishment of a Nonessential Experimental Population of Grizzly Bears in the Bitterroot Area of Idaho and Montana*. Denver CO: U.S. Department of the Interior, Fish and Wildlife Service, 2000.

———. *Grizzly Bear Recovery in the Bitterroot Ecosystem: Summary of Draft Environmental Impact Statement*. Washington DC: U.S. Government Printing Office, 1997.

U.S. National Archives and Record Administration. "Amendment Listing the Grizzly Bear of the 48 Conterminous States as a Threatened Species." *Federal Register* 40, no 145 (July 28, 1975).

———. "Reintroduction of Grizzly Bears to the Bitterroot Ecosystem of East-Central Idaho and Western Montana." *Federal Register* 60, no. 5 (January 9, 1995).

Velado, Carlos L. "Grizzly Bear Reintroduction to the Bitterroot Ecosystem: Perceptions of Individuals with Land-based Occupations." Master's thesis, University of Montana, 2005.

Weaver, John. "Of Wolves and Bears." *Western Wildlands*, Fall 1986.

Werblow, Steve. "Timber & Wildlife Bear with Each Other." *Future Earth* 2, no. 1 (1996).

Wiebe, Robert H. *Self-Rule: A Cultural History of American Democracy*. Chicago: University of Chicago Press, 1995.

Wilkinson, Charles. "Paradise Revised." In *Atlas of the New West: Portrait of a Changing Region*, edited by James Robb, William E. Riebsame, and Hannah Gosnell. New York: W. W. Norton, 1997.

Wiltsie, Meredith, and Michael Wyckoff. "Reinventing Red Lodge: The Making of a New Landscape, 1884–2000." In *Imagining the Big Open: Nature, Identity, and Play in the West*, edited by Liza Nicholas, Elaine M. Bapis, and Thomas J. Harvey, 125–50. Salt Lake City: University of Utah Press, 2003.

Winkler, Richelle, Donald R. Field, A. E. Luloff, Richard S. Krannich, and Tracy Williams. "Social Landscapes of the Inter-Mountain West: A Comparison of 'Old West' and 'New West' Communities." *Rural Sociology* 72, no. 3 (2007): 478–501.

Wrobel, David. *The End of American Exceptionalism: Frontier Anxiety from the Old West to the New Deal*. Lawrence: University Press of Kansas, 1993.

Wright, William H. *The Grizzly Bear: The Narrative of a Hunter-Naturalist.* Lincoln: University of Nebraska Press, 1977. First published 1909 by T. Werner Laurie, London.

Yaffee, Steven Lewis. *The Wisdom of the Spotted Owl: Policy Lessons for a New Century.* Washington DC: Island Press, 1994.

Index

Page numbers in italics signify graphics. Photos follow p. 136.

www.ingramcontent.com/pod-product-compliance
Lightning Source LLC
Chambersburg PA
CBHW030422100426
42812CB00028B/3067/J